智能微型运动装置（Micromouse）技术与应用系列丛书
天津市"一带一路"联合实验室（研究中心）项目研究成果
工程实践创新项目（EPIP）教学模式规划教材

智能鼠原理与制作
（进阶篇）

王 超　高 艺　宋立红　编著
王 娟　范平平　严靖怡　编译

中国铁道出版社有限公司
CHINA RAILWAY PUBLISHING HOUSE CO., LTD.

内 容 简 介

本书为中英双语版，以天津启诚伟业科技有限公司提供的TQD-Micromouse-JD智能鼠为载体，是智能微型运动装置（Micromouse）技术与应用系列丛书的进阶篇。

本书以真实工程项目为背景，通过"基础知识篇"、"综合实践篇"和"拓展竞技篇"三篇讲述了智能鼠的发展、硬件、开发环境、基本操作；智能鼠高级功能、智能鼠实战任务；智能鼠的路径规划和行为决策算法、智能鼠路径规划原理、智能鼠走迷宫程序设计等。同时，本书附录提供了国际Micromouse 走迷宫竞赛相关知识、智能鼠迷宫图库、专业词汇中英文对照表、国际实训课程标准等丰富资源。

本书在重要的知识点、能力点和素养点上，配有丰富的视频、图片、文本等资源，学习者可以通过扫描书中二维码获取相关信息。

本书适合作为职业院校相应专业综合与创新实践教学的教材，还可作为相关工程技术人员培训用书及智能鼠爱好者参考用书。

图书在版编目（CIP）数据

智能鼠原理与制作.进阶篇:汉、英/王超,高艺,宋立红编著.—北京:中国铁道出版社有限公司,2021.2
（智能微型运动装置（Micromouse）技术与应用系列丛书）
ISBN 978-7-113-27538-9

Ⅰ.①智… Ⅱ.①王… ②高… ③宋… Ⅲ.①智能机器人-程序设计-汉、英 Ⅳ.① TP242.6

中国版本图书馆 CIP 数据核字 (2020) 第 273224 号

书　　名：	智能鼠原理与制作（进阶篇）
作　　者：	王　超　高　艺　宋立红

策　　划：何红艳	编辑部电话：（010）83552550
责任编辑：何红艳　绳　超	
封面设计：刘　颖	
责任校对：孙　玫	
责任印制：樊启鹏	

出版发行：中国铁道出版社有限公司（100054，北京市西城区右安门西街8号）
网　　址：http://www.tdpress.com/51eds/
印　　刷：三河市兴博印务有限公司
版　　次：2021年2月第1版　2021年2月第1次印刷
开　　本：787 mm×1 092 mm　1/16　印张：14.75　字数：248千
书　　号：ISBN 978-7-113-27538-9
定　　价：56.00元

版权所有　侵权必究

凡购买铁道版图书，如有印制质量问题，请与本社教材图书营销部联系调换。电话：（010）63550836
打击盗版举报电话：（010）63549461

作者简介

王 超

天津大学电气自动化与信息工程学院教授,教育部高等学校自动化类专业教学指导委员会委员,主要从事多相流检测与仪器和电学层析成像的研究(ERT, ECT, EMT 和 EST),在天津大学教授计算控制技术和工业控制网络课程,自 2010 年起,首次将智能鼠作为重要的实践教学载体引入电气自动化与信息工程学院。2018 年,在第 33 届 APEC 国际电脑鼠竞赛中,天津大学的两支队伍包揽冠亚军。

高 艺

南开大学电子信息工程学院硕士生导师,电子信息实验教学中心副主任,天津市单片机学会青年骨干工作委员会副主任,多项天津市大学生竞赛及职业技能裁判组成员。先后参与多项"国家高技术研究发展计划(863 计划)项目"、"天津市科技支撑计划重点项目"以及横向科研项目。多次作为指导教师带队参加全国大学生电子设计竞赛、天津市电子设计竞赛、天津市物联网竞赛、天津市大学生 IEEE 电脑鼠竞赛、APEC 国际电脑鼠竞赛、全国机器人大赛等。

宋立红

天津启诚伟业科技有限公司总经理,启诚智能鼠创始人。多年来专注致力于高等教育、职业教育、基础教育领域的嵌入式、物联网、人工智能等教学仪器设备研发、设计、生产、推广、服务工作。40 余次赞助支持大学生学科竞赛"启诚杯"智能鼠走迷宫赛项及职业院校技能竞赛智能微型运动装置(智能鼠)赛项等。从 2016 年开始,积极致力于国际项目鲁班工坊技术支持服务工作,启诚智能鼠作为中国创新型教育装备,伴随鲁班工坊不远万里前往泰国、印度、印尼、巴基斯坦、柬埔寨、尼日利亚、埃及等国家,受到所在国师生一致青睐,为"一带一路"倡议做出了努力和贡献。

王 娟

天津轻工职业技术学院国际交流处副处长、副教授。自 2016 年作为主要参与者，完成印度、埃及鲁班工坊建设。曾指导学生参加天津市英语口语大赛获一等奖，参加天津海河教育园首届高职英语口语比赛获三等奖，参加（新加坡）全球品牌策划大赛中国地区选拔赛获优秀奖。参与教学成果奖项目，获国家级教学成果一等奖，天津市教学成果奖特等奖和二等奖各一项，2017 年至今主持及参与完成多项课题。参与翻译了国际化教材《数控机床装调与检测》。

范平平

天津轻工职业技术学院电子信息与自动化学院讲师，电气自动化技术专业教研室主任。参加全国职业院校信息化教学设计比赛获二等奖，参加全国职业院校教师微课大赛获二等奖。天津市高职高专院校学生技能大赛优秀指导教师，全国职业院校技能大赛优秀指导教师。指导学生参加"智能鼠走迷宫"、"风光互补发电系统安装与调试"、"大学生机器人大赛"、"天津市大学生创新方法应用大赛"和"智能电梯装调与维护"赛项并获奖几十项。国家级新能源类专业教学资源库课程"单片机控制技术"主讲教师，天津市高校新时代"课程思政"改革精品课"单片机应用技术"主讲教师。

严靖怡

天津启诚伟业科技有限公司总经理助理，就读于美国加州大学圣克鲁兹分校。2015 年担任美国 APEC 智能鼠国际竞赛组委会主席 MIT David Otten 教授在中国交流访问期间随行翻译。2018 年担任新蒙古教育集团董事会副主席 Davaanyam 访问天津智能鼠竞赛交流活动随行英语翻译及会议同声传译。2018 年自费到柬埔寨做志愿者，担任柬埔寨国立理工学院 Bun Phearin 校长鲁班工坊智能鼠项目翻译，帮助柬埔寨学生学习中国智能鼠技术，为"一带一路"沿线国家教育发展做出了努力和贡献。

扫码观看

智能鼠名人榜

——Mr. David Michael Otten

I am very pleased to find out that you are going to write a book about micromouse. This contest is a fantastic way to learn about electromechanical systems and integrating hardware and software. I have learned a great deal during my 30 years with this contest and I am sure your readers will also. Congratulations and good luck with this endeavor.

David Otten
APEC micromouse chair

我很高兴得知你们要编写关于智能鼠的书。智能鼠大赛是学习机电系统和集成软硬件的绝佳方式。在过去经历的30年比赛中，我学到了很多东西，我相信你们的读者也会这样。祝贺你们并祝福好运！

Mr. David Michael Otten

美国麻省理工学院 高级研发工程师。
国际智能鼠走迷宫教育教学专家。
多年从事AI机器人开发研究工作，连续30余届美国APEC国际电脑鼠竞赛组委会主席，曾多次参加日本、新加坡、美国、英国等国家智能鼠比赛，并多次蝉联世界冠军。

扫码观看 智能鼠名人榜 ——Mr. Peter Harrison

The micromouse contest is an integration of multiple disciplines and many technologies. It involves machine engineering, electronic engineering, automatic control, artificial intelligence, program design, sensing and testing technology.

The micromouse contest will enhance the participant's technology level and application abilities, providing a platform for technological innovation

The publication of books on micromouse education will play a significant role in learning micromouse technology for Chinese students. The micromouse robots made by Qicheng are to the world through Luban workshops, benifitting students around the world

Congratulations on the publication of the micromouse book series. They provide convenience and reference for micromouse fans and students at all levels.

  智能鼠是集多学科多技术的融合体,主要涉及机械工程、电子工程、自动控制、人工智能、程序设计、传感与测试技术等学科。
  竞赛的开展,必然提升参赛者在相关领域的技术水平和应用能力,为技术创新提供平台。
  这类教育资源书籍的出版,对于中国乃至世界学生学习智能鼠技术都有着重要的意义和作用,启诚电脑鼠通过鲁班工坊冲出国门走向世界,让世界各地的学生受益,我为启诚智能鼠感到骄傲和自豪!
  由衷地祝贺智能微型运动装置(Micromouse)技术与应用系列丛书出版。这为智能鼠爱好者、不同教育层面的学生学习嵌入式微型机器人(智能鼠)提供了便利和参考。

Mr. Peter Harrison

  英国伯明翰城市大学 高级研发工程师。
  国际智能鼠走迷宫教育教学专家。
  多年来从事设计、研发IT集成项目工作,在培养大学生人工智能机器人技术领域、实训教学和实用技能方面成绩卓著,曾多次参加美国、日本、新加坡、英国等国家的智能鼠走迷宫比赛,并多次蝉联世界冠军。

扫码观看

智能鼠名人榜

——Mr. António Valente

> When about ten years ago I started organizing robotics events with the Micromouse contest, I did it because I realized that this robotics contest would be the most applicable to encourage students to STEM areas. It is a complete contest that involves students of all ages and skill levels. However, there is still very little aggregated information on this topic. Thus, the publication of a series of books on Micromouse is beneficial.
>
> It is with great pleasure that I, as an enthusiast and as the organizer of the Micromouse Portuguese Contest, give my support to this series of publications on Micromouse, being certain that they will contribute to the enhancement of Micromouse.

十年前,我开始组织智能鼠比赛,因为我认为这个比赛最适合激发学习者对 STEM 领域的热情。各个年龄和各种水平的选手都可以参加这个比赛。但遗憾的是,有关这个领域鲜有详细的介绍。因此,有关智能鼠的这套书十分有益。作为一个智能鼠爱好者和葡萄牙智能鼠赛事的组织者,我向大家推荐这套书。我认为这套书将会促进智能鼠的发展。

Mr. António Valente

葡萄牙 Trás-os-Montes and Alto Douro 大学科学技术学院教授,高级研究员。

葡萄牙国际智能鼠大赛组委会主席。

研究方向:MEMS 传感器、微控制器和嵌入式系统,重点是农业应用。完成了多项葡萄牙国家级和国际级资助的纵向和横向科研项目和课题,包括:安全、Eno-分析、RobTech、IPAVPSI、Focus 等。

前言

"智能鼠",英文名为Micromouse,是使用嵌入式微控制器、传感器和机电运动部件构成的一种智能微型运动装置(嵌入式微型机器人),智能鼠可以在不同"迷宫"中自动记忆和选择路径,采用相应的算法,快速到达所设定的目的地。智能鼠走迷宫竞赛结合了机械电子、控制、光学、程序设计和人工智能等多方面的科技知识。

四十多年来,电气电子工程师学会(IEEE)每年举办一次国际性的智能鼠走迷宫竞赛,自举办以来各国踊跃参加,尤其是美国和欧洲国家的高校学生,为此有些大学还特别开设了"智能鼠原理与制作"的选修课程。中国从2007年开始在上海长三角地区举行小规模尝试性比赛。2009年,天津启诚伟业科技有限公司将这项国际赛事引进天津,以工程实践创新项目(EPIP)教学模式,对智能鼠走迷宫竞赛进行本土化创新改革,对于后期智能鼠竞赛的开展和走进课堂、融入教学起到关键性的作用。经过多年的蜕变与优化,"智能鼠"已经成为集"职业性、综合性、先进性、趣味性"于一体的创新实践教育平台,在推动课程改革、提高教学质量、培养学习者的工程实践创新能力等方面发挥了重要的作用。

为了将智能鼠的成果进一步推广应用,我们编写了适用于职业院校学生学习及培训的《智能鼠原理与制作》(进阶篇)教材。本书以天津启诚伟业科技有限公司提供的TQD-Micromouse-JD智能鼠为载体,由浅及深、由易到难地进行实践教学。

本书遵循递进原则,从"玩转"到"掌握",再到"精通",丰富学习者的工程实践知识和技术应用经验,拓展学习者的专业视野,内化形成良好的职业素养,提升学习者的实践创新能力。本书所选案例均来自真实的工程项目,编者均来自国内长期从事智能鼠研究与开发、国际智能鼠走迷宫竞赛获奖的院校和企业。

本书在重要的知识点、能力点和素养点上,配有丰富视频、图片、文本等资源,学习者可以通过扫描书中二维码获取相关信息。本书编著者长期的国际化教学活动积淀,使得本书成为推进国际化人才培养的实践教学载体,

本书适合作为职业院校相应专业综合与创新实践教学的教材，还可作为相关工程技术人员培训用书及智能鼠爱好者参考用书。

在本书附录中提供了"智能鼠原理与制作"（适用于高等职业学校）国际实训课程标准。明确了以智能鼠的硬件结构、智能鼠的开发环境、智能鼠的红外检测、智能鼠的运动姿态控制、智能鼠的路径规划和行为决策法则等项目为专业核心能力。采用"教、学、做"一体化的方式，完成专业能力、社会能力、方法能力的培养。本课程内容与多个国家的"鲁班工坊"建设项目高度融合。服务"一带一路"倡议，推广中国职业教育标准，为"一带一路"沿线国家提供丰富实践教学资源，服务各地技术技能人才培养。智能微型运动装置（Micromouse）技术与应用系列丛书是天津市"一带一路"联合实验室（研究中心）——天津中德柬埔寨智能运动装置与互联通信技术推广中心研究成果，同时也是工程实践创新项目（EPIP）教学模式规划教材。在高等教育领域"新工科"方向，可使用本书作为信息与自动化技术融合与创新学科教学的指导用书。

本书由天津大学教授王超，南开大学副教授高艺，启诚智能鼠创始人、天津启诚伟业科技有限公司总经理宋立红编著。英文部分由天津轻工职业技术学院副教授王娟、讲师范平平，天津启诚伟业科技有限公司总经理助理严靖怡编译。天津交通职业学院副教授刘宝生、讲师李萌以及天津轻工职业技术学院讲师王丹阳参与编写。国际专家美国麻省理工学院教授David Otten和英国伯明翰城市大学教授Peter Harrison、葡萄牙Trás-os-蒙特斯与奥拓杜罗大学教授António Valente参与本书英文内容的译审，并专为本书写了贺信。本书在编写过程中得到了天津大学、南开大学、天津轻工职业技术学院、天津交通职业技术学院、美国麻省理工学院、英国伯明翰城市大学和葡萄牙Trás-os-蒙特斯与奥拓杜罗大学等相关院校教授、专家的大力支持。天津启诚伟业科技有限公司陈立考、邱建国、宋姗为本书出版提供了企业实际工程案例、二维码视频、动画、PPT等课程资源。衷心感谢天津市教育委员会、中国铁道出版社有限公司、天津启诚伟业科技有限公司对本教学资源开发提供的指导与帮助。本书由天津轻工职业技术学院赞助编译，中国铁道出版社有限公司支持出版，并通过鲁班工坊在"一带一路"沿线国家使用。

限于编著者的经验、水平以及时间，书中难免存在不妥和缺漏，敬请专家、广大读者批评指正。

编著者

2020年8月

目录

第一篇 基础知识篇 1

项目一 智能鼠的发展历程 003
- 任务一 智能鼠的起源 003
- 任务二 智能鼠的竞赛与调试环境 009

项目二 智能鼠的硬件结构 012
- 任务一 智能鼠的组成 012
- 任务二 智能鼠的核心控制板 013

项目三 智能鼠的开发环境 015
- 任务一 IAR EWARM开发环境 015
- 任务二 智能鼠的程序下载 016

项目四 智能鼠的基本功能操作 017
- 任务一 智能鼠的人机交互系统 017
- 任务二 智能鼠的红外检测 020
- 任务三 智能鼠的运动姿态控制 025

第二篇 综合实践篇 035

项目一 智能鼠高级功能 037
- 任务一 智能鼠直行的控制方法 037
- 任务二 智能鼠精确转弯的控制方法 042

项目二 智能鼠实战任务 046
- 任务一 7289调试模块的使用 046
- 任务二 非接触式启动和停止控制 048
- 任务三 "8字型"路径运行控制 050

第三篇 拓展竞技篇 053

项目一 智能鼠的路径规划和行为决策算法 055
- 任务一 迷宫搜索的常用策略 055

任务二　迷宫搜索的基本法则 055
　项目二　智能鼠路径规划的原理 060
　　　任务一　迷宫信息的存储方法 060
　　　任务二　等高图的制作方法 061
　项目三　智能鼠走迷宫程序设计 062
　　　任务一　运行姿态程序控制 062
　　　任务二　基本程序结构解析 063

附录 077

　附录A　风靡全球的国际智能鼠走迷宫竞赛 077
　附录B　进阶级经典竞赛案例分析 086
　附录C　TQD-Micromouse-JD器件清单 092
　附录D　教学内容和学时安排 092
　附录E　电路图形符号对照表 093
　附录F　专业词汇中英文对照表 094
　附录G　"智能鼠原理与制作"国际实训课程标准 096

第一篇 基础知识篇

　　智能鼠走迷宫竞赛在国际上已经有40多年的历史，竞赛要求智能鼠从起点出发，在不受人为操纵影响的条件下在未知的迷宫中，自主搜索迷宫找到终点，并挑选出最短的一条路径进行冲刺。竞赛根据搜索迷宫的时间和冲刺到终点所用时间分出名次。竞赛迷宫遵照电气电子工程师学会（IEEE）的国际标准。在本篇中，将分别从国际IEEE标准迷宫场地、智能鼠的硬件系统和软件开发环境等方面系统介绍智能鼠技术，并对智能鼠的基本原理和实际操作方法进行具体说明。

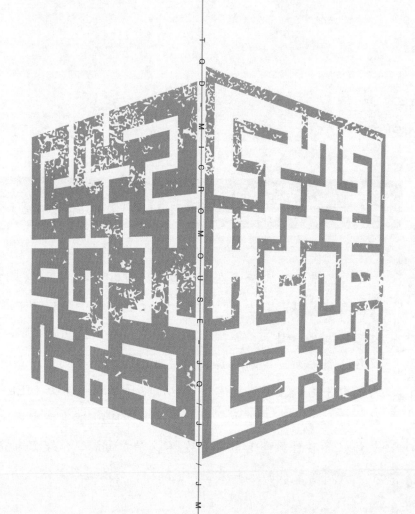

项目一 智能鼠的发展历程

学习目标

（1）了解智能鼠的发展历程。
（2）理解智能鼠走迷宫竞赛平台——竞赛迷宫场地、全自动计分系统。

任务一 智能鼠的起源

一、智能鼠的起源

1938年，美国密歇根州的数学家香农（Claude Elwood Shannon）完成了《继电器和开关电路的符号分析》的论文。由于布尔代数只有0和1，恰好与二进制对应，香农将它运用于以脉冲方式处理信息的继电器开关，从理论到技术彻底改变了数字电路的设计方向，因此，这篇论文在现代数字计算机史上具有划时代的意义。

1948年，香农又发表了一篇至今还在闪烁光芒的论文——《通信的数学理论》，从而给自己赢得了"信息论之父"的桂冠。

1956年，他参与发起了达特默斯人工智能会议，成为这一新学科的开山鼻祖之一。他不仅率先把人工智能运用于计算机下棋方面，而且还发明了一个能自动穿越迷宫的"智能鼠"，以此证明计算机可以通过学习提高智能。

二、智能鼠的国际发展历程

1972年，《机械设计》杂志发起了一场竞赛。在竞赛中，仅由捕鼠器弹簧驱动的机械鼠，不停地与其他参赛鼠竞赛，以判断哪个机械鼠能够沿着跑道跑出最长的距离。

1977年，IEEE Spectrum 杂志提出智能鼠的观念。智能鼠是一个小型的由微处理器控制的机器人车辆，在复杂迷宫中具有译码和导航的功能和能力。

1979年，电气电子工程师学会（IEEE）通过其Spectrum and Computer杂志

发起了一场智能鼠竞赛，奖励能够在最短时间内自主走出迷宫的智能鼠的设计者1 000美元。

1980年，东京举办了首场全日本Micromouse国际公开赛，之后，又有多个比赛被创办，如：1980年英国智能鼠大赛，1987年新加坡举办了第一届新加坡Micromouse竞赛和2007年中国计算机学会举办的首场IEEE国际标准Micromouse走迷宫竞赛等，如图1-1-1所示。

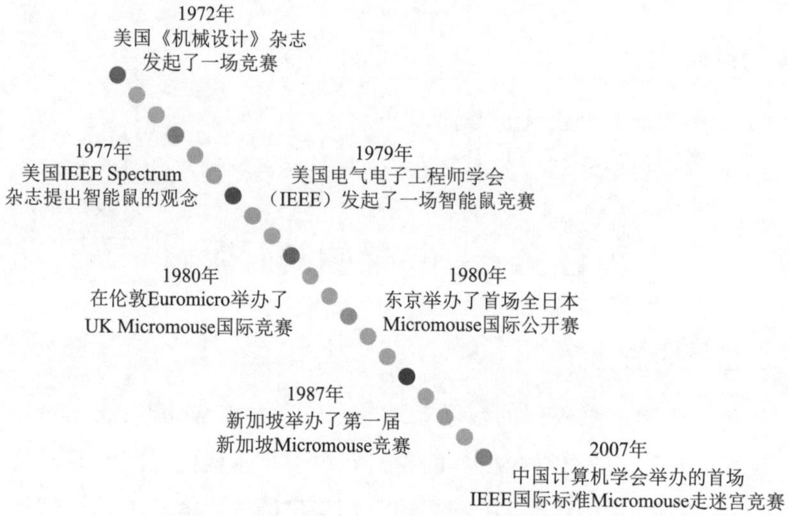

1972年
美国《机械设计》杂志
发起了一场竞赛

1977年
美国IEEE Spectrum
杂志提出智能鼠的观念

1979年
美国电气电子工程师学会
（IEEE）发起了一场智能鼠竞赛

1980年
在伦敦Euromicro举办了
UK Micromouse国际竞赛

1980年
东京举办了首场全日本
Micromouse国际公开赛

1987年
新加坡举办了第一届
新加坡Micromouse竞赛

2007年
中国计算机学会举办的首场
IEEE国际标准Micromouse走迷宫竞赛

图1-1-1　智能鼠国际发展

从最初1972年的机械电子鼠发展到现在的智能鼠，经过了40多年沧海桑田的蜕变，参加竞赛的选手从开始仅限于哈佛大学、麻省理工学院等世界知名学府的研究生，发展到从研究型大学到应用技术大学再到职业院校的学生，甚至是中小学生。多教育层面都采纳了智能鼠为教学载体，培养学生们的工程素养以及科技创新意识、动手设计能力。

各类智能鼠竞赛也如雨后春笋般蓬勃发展。目前智能鼠竞赛已经成为应用于不同教育阶段的国际创新型学生竞赛。

三、智能鼠的中国发展历程

从2007年至今，智能鼠在中国经历了十余年的发展历程，如图1-1-2所示。2007年天津启诚伟业科技有限公司将这项国际赛事引进天津，以中国先进的教育模式"工程实践创新项目"为核心理念，对智能鼠走迷宫竞赛进行本土化创新改革，助力智能鼠竞赛在中国的蓬勃开展，对于智能鼠技术走进课堂融入教学起到关键性的引领作用。

● 视频
中国智能鼠发展

```
首场比赛        高等教育        职业教育        普通教育        国际竞赛
  ●              ●              ●              ●              ●
2007年        2009年至今      2010年至今      2016年至今      2016年至今
中国首场智能  大学生学科竞赛  职业院校学生    普职融通国际    中国IEEE智能鼠
鼠走迷宫竞赛  Micromouse大赛  技能竞赛        挑战赛          国际邀请赛
                              Micromouse大赛  Micromouse大赛
```

图1-1-2 智能鼠在中国的发展

竞赛对于满足产业优化升级，开阔国际视野，掌握实践与创新经验，培养高技术、高技能人才，起到了引领推动作用（见图1-1-3）。智能鼠在中国从大学生竞赛到职业院校大赛，再到普职融通国际挑战赛，积累了丰富的竞赛经验和优秀的技术积淀。

图1-1-3 竞赛纪实照片

十余年来，中国的智能鼠竞赛不断创新国际发展新思路，从最初的"简单模仿"学习，发展到目前的"互学互鉴"，逐步搭建起国际交流合作的新平台，先后经历了学习借鉴、蜕变升华和引领辐射三个阶段。

首先是学习借鉴：2015年天津大学生代表队征战美国第30届APEC世界Micromouse竞赛（见图1-1-4），取得了世界第六的好成绩。2017年至2018年，天津启诚伟业科技有限公司全额资助了在天津大学生智能鼠竞赛上获得企业命题赛冠军队，到日本东京参加第38届和第39届全日本Micromouse国际公开赛（见图1-1-5），学习借鉴国际智能鼠先进技术，结识众多智能鼠业界专家教授，对中国智能鼠技术的发展与提升起到推动的作用。

接着是蜕变升华：智能鼠大赛在中国进行本土化创新改革，设计了一系列从易到难的启诚智能鼠教学平台，满足"中、高、本、硕"不同学习阶段学生学习应用。从2016年开始先后邀请美国麻省理工学院的David Otten教授、中国台湾龙华科技大学苏景晖教授、新加坡义安理工学院黄明吉教授、英国伯明翰城市大学Peter Harrison教授、日本智能鼠国际公开赛组委会秘书长中川友纪子先生等智能鼠专家和来自泰国、印度、印尼、巴基斯坦、柬埔寨等国际"鲁

视频
学习借鉴
（美国APEC）

视频
蜕变升华
（第一届IEEE）

班工坊"师生,以及来自中国天津、北京、河南、河北等国内省市精英级代表队,先后加盟中国IEEE智能鼠走迷宫国际邀请赛(见图1-1-6、见图1-1-7)。国际选手通过参加中国比赛,对中国竞赛标准、竞赛规则、竞赛模式和竞赛理念有了更深层次的了解和认同,从而切实推动了国际化的交流与合作,达到"互学互鉴"的目的。

图1-1-4　中国天津代表队远征美国参加国际大赛

图1-1-5　中国天津代表队远赴日本参加国际大赛

图1-1-6　第三届IEEE智能鼠走迷宫国际邀请赛

图1-1-7 "启诚杯"第四届IEEE智能鼠走迷宫国际邀请赛

最后是引领辐射：教育对外开放是我国改革开放事业的重要组成部分，随着"一带一路"倡议的推进，2016年以来在中国教育部指导下，先后启动了海外鲁班工坊国际项目，智能鼠作为中国优秀的教育装备，伴随着鲁班工坊走出国门与世界分享。从2016年至今，启诚智能鼠来到泰国、印度、印尼、巴基斯坦、柬埔寨、尼日利亚、埃及等海外国家，免费开展智能鼠竞赛的推广和课程培训，受到了沿线国家师生的一致青睐（见图1-1-8~图1-1-13）。智能鼠成为连接世界的纽带与桥梁！

视频

引领辐射
（印度鲁班）

图1-1-8 印度鲁班工坊开展智能鼠培训课程

图1-1-9 2016年泰国鲁班工坊开展智能鼠培训课程

图1-1-10　2017年印尼鲁班工坊开展智能鼠培训课程

图1-1-11　2018年巴基斯坦鲁班工坊开展智能鼠培训课程

图1-1-12　2018年柬埔寨鲁班工坊开展智能鼠培训课程

图1-1-13　2020年埃及鲁班工坊开展智能鼠培训课程

任务二 智能鼠的竞赛与调试环境

一、竞赛迷宫场地

目前，国际和国内比赛都使用同样规格的比赛场地，即一个由8×8个格子组成的方形迷宫。迷宫的"墙壁"是可以插拔的，这样就可以形成各种各样的迷宫。

如图1-1-14所示为TQD-Micromouse Maze 8×8比赛场地。迷宫底板的尺寸为2.96 m×2.96 m，上面共有 8×8 个标准迷宫单元格。图1-1-15所示为古典智能鼠迷宫挡板和立柱。

图1-1-14　TQD-Micromouse Maze 8×8

图1-1-15　古典智能鼠迷宫挡板和立柱

TQD-Micromouse Maze 8×8迷宫场地规范如下：

（1）迷宫由8×8个、18 cm×18 cm大小的正方形单元所组成。

（2）迷宫的挡板高5 cm，厚1.2 cm，因此两个挡板所构成的通道的实际距离为16.8 cm，挡板将整个迷宫封闭。

（3）迷宫挡板的侧面为白色，顶部为红色。迷宫的地面为木质，颜色为哑光黑。挡板侧面和顶部的涂料能够反射红外线，地板能够吸收红外线。

（4）迷宫的起始单元可选设在迷宫四个角落之中的任何一个。起始单元必须三面有挡板，只留一个出口。迷宫的终点设在迷宫中央，由四个正方形单元构成。

（5）在每个单元的四角可以插上一个小立柱，其截面为正方形。如图1-1-14所示。立柱长1.2 cm、宽1.2 cm、高5 cm。小立柱所处的位置称为"格点"。除了终点区域的格点外，每个格点至少要与一面挡板相接触。

（6）迷宫制作的尺寸精度误差应不大于5%，或小于2 cm。迷宫地板的接缝不能大于0.5 mm，接合点的坡度变化不超过4°。挡板和立柱之间的空隙不大于1 mm。

（7）起点和终点设计遵照IEEE智能鼠竞赛标准，即智能鼠按照顺时针方向开始运行。

二、专用测试场地

专用测试场地上绘有13个标记位置,并且使用不同的颜色进行区分(见图1-1-16),用于调试红外传感器和优化转弯控制参数。接下来就带领读者认识一下它:

(1)①至②,灰色通道,用来检测智能鼠在无红外校准的情况下直行的偏移量。

(2)③深红色矩形,④橙色矩形;③至②、④至②均是用来检验有红外校准时的智能鼠直行情况。

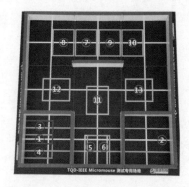

图1-1-16 TQD-IEEE Micromouse 专用测试场地

(3)⑤黄色矩形用来调节智能鼠左前红外强度,⑥绿色矩形用来调节智能鼠右前红外强度;校正车姿。

(4)⑦、⑧绿色矩形用来调节智能鼠右后红外强度,⑨、⑩绿色矩形用来调节智能鼠左后红外强度;检测路口。

(5)⑪、⑫、⑬三个蓝色矩形用来调试智能鼠转弯90°。

三、全自动计分系统

为了精确计量智能鼠完成迷宫的时间,需要全自动地计算智能鼠通过起点和终点的时间。图1-1-17所示为由天津启诚伟业科技有限公司设计生产的用于智能鼠走迷宫竞赛的电子自动计分系统。

TQD-Micromouse Timer V2.0系统包含起点对射模块、终点对射模块、智能鼠计分系统模块、计分软件等。

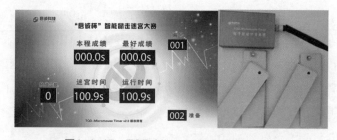

图1-1-17 TQD-Micromouse Timer V2.0

起点对射模块和终点对射模块采用迷你USB充电方式,通过内置的一组激光对射传感器检测智能鼠经过。智能鼠计分系统模块用于接收起点对射模块和终点对射模块通过ZigBee发过来的数据,经过计算机中的计分软件处理,以一

种直观的方式展现智能鼠在迷宫中的运行情况。计分软件也可以单独使用，可通过鼠标输入起点事件和终点事件。计分系统整体的计时精度可达0.001 s。

起点对射模块和终点对射模块分别安装在起点迷宫格和终点迷宫格中，如图1-1-18、图1-1-19所示。当智能鼠经过时，激光被阻断，从而产生起点或终点信号。

图1-1-18　迷宫起点

图1-1-19　迷宫终点

思考与总结

（1）IEEE国际标准智能鼠场地由哪些部分组成？

（2）请归纳智能鼠竞赛的特点。

（3）全自动计分系统大大提高了竞赛成绩计算的准确性，请简要说明其工作原理。

项目二 智能鼠的硬件结构

学习目标

（1）认识智能鼠的基本硬件结构。

（2）学习智能鼠中央处理器的运用。

TQD-Micromouse-JD（见图1-2-1）实训教学型智能鼠是天津启诚伟业科技有限公司根据教学需求，自主研发设计生产的教学竞赛型智能机器人。智能鼠可用于大专院校智能鼠实训教学课程，亦可作为入门级IEEE国际标准智能鼠走迷宫竞赛平台，更是各类智能型机器人竞赛的首选。

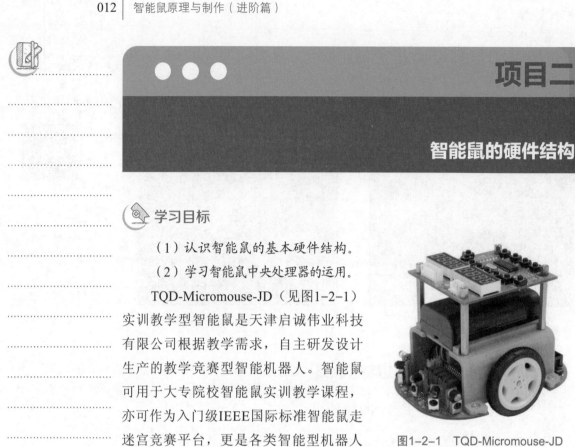

图1-2-1　TQD-Micromouse-JD

任务一　智能鼠的组成

TQD-Micromouse-JD智能鼠硬件组成及其优点：

（1）微处理器采用LM3S615亦或STM32，基于ARM Cortex-M3内核。具有运算速度快、中断响应快、外设丰富等优点，保证了智能鼠的高度智能化。另外，提供了丰富的函数库，只要懂C语言就能开发，大大降低了智能鼠使用难度。

（2）智能鼠身长12 cm，宽9 cm，短小精悍，机械结构简单，安装方便，五组红外数字传感器实时检测迷宫挡板信息，可灵活的在迷宫格中完成90°和180°转弯。

（3）智能鼠采用双步进电动机驱动，运行平稳，无须减速装置，机械结构简单，特别适合于初学者学习和应用。

（4）CPU控制板设置Start键、复位键和一个10针JTAG调试接口，并预留

视　频

智能鼠组成

六个GPIO、一个串口和一个SPI接口，方便用户自由扩展。

（5）配套有标准充电器以及2 200 mA·h，7.4 V的可充电锂电池。

（6）LM LINK USB JTAG 调试器使用简单，支持在线调试，为学生进行程序设计提供最大方便。

TQD-Micromouse-JD智能鼠硬件结构，如图1-2-2所示。

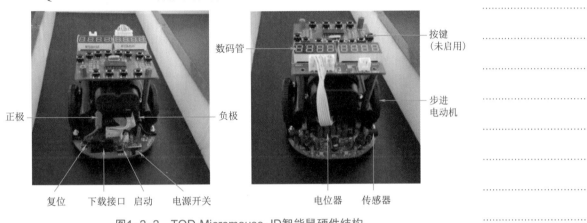

图1-2-2　TQD-Micromouse-JD智能鼠硬件结构

TQD-Micromouse-JD智能鼠的电路组成框图如图1-2-3所示，主要由核心控制模块、输入模块和输出模块构成，其中核心控制模块主要包括核心板电路、电源电路和控制电路，此外由核心板电路扩展出部分外围电路构成输入/输出模块，在智能鼠运行过程中发挥信号传送和接收功能。主要包括键盘显示电路、JTAG接口电路、电动机驱动电路、按键电路、红外检测电路。

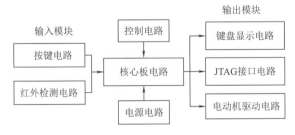

图1-2-3　TQD-Micromouse-JD智能鼠的电路组成框图

任务二　智能鼠的核心控制板

智能鼠核心控制板是整个智能鼠控制的核心，相当于智能鼠的CPU。TQD-Micromouse-JD智能鼠采用一片QFP-48N封装的LM3S615和极少的外围器件构成。

核心控制板电路如图1-2-4所示，LM3S615是核心板电路的主体，LM3S615之外的电路为集成芯片的外围控制电路。其中，晶振Y1、电容C27和C28构成脉冲振荡电路，向微控制器的引脚9和引脚10发送脉冲信号。

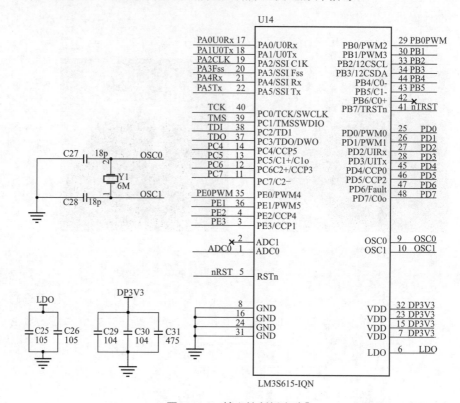

图1-2-4　核心控制板电路[①]

思考与总结

（1）智能鼠各个模块之间是如何进行数据传输的？

（2）智能鼠各部分有无相似的电子元器件可以替代？

（3）TQD-Micromouse-JD智能鼠由传感器、控制器和执行器三部分组成。红外线传感器检测周围障碍物距离，并由数码管直观显示出来；控制器依据这些红外数据控制步进电动机进行动作，从而实现避障。

---

① 类似图稿为Protel 99SE导出的原理图，其图形符号与国家标准符号不一致，二者对照关系参见附录E。

项目三

智能鼠的开发环境

📖 **学习目标**

（1）学习IAR EWARM安装。
（2）学习智能鼠的程序下载。

任务一　IAR EWARM开发环境

TQD-Micromouse-JD采用IAR Embedded Workbench for ARM（以下简称IAR）作为程序开发环境。它包含项目管理器、编辑器、C/C++编译器和ARM汇编器、连接器XLINK和支持RTOS的调试工具C-SPY。在EWARM环境下可以使用C/C++方便地开发嵌入式应用程序。比较其他的ARM开发环境，IAR EWARM具有入门容易、使用方便和代码紧凑等特点。

我们提供完整的驱动库和全迷宫Demo例程，包括底层驱动、顶层智能算法以及基础实验程序。读者只要懂C语言就能开发。

软件界面如图1-3-1所示。

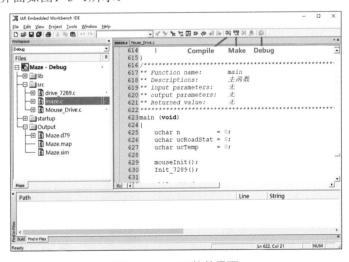

图1-3-1　IAR软件界面

软件

智能鼠IAR开发环境下载

任务二　智能鼠的程序下载

一、Luminary驱动库

TQD-Micromouse-JD主控芯片采用LM3S615，Demo程序中使用了Luminary驱动库，所以在下载程序前需要在软件的安装目录中添加Luminary驱动库。

二、J-Link下载器

J-Link（见图1-3-2）专门用于对Luminary系列单片机程序的调试与下载。该下载器结合IAR EWARM集成开发环境，可支持所有LM3S系列MCU的程序的下载与调试。

J-Link采用USB接口与计算机连接，无论是台式计算机还是笔记本式计算机都应用自如。

图1-3-2　J-Link

三、连接硬件并下载程序

图1-3-3　硬件连接

下载程序前一定要正确连接智能鼠、下载器和计算机（见图1-3-3）。J-Link下载器的右上角是VCC，智能鼠插座的右上角同样是VCC，使用专用下载线对应连接即可。

四、单击IAR软件中的Debug按钮下载程序

思考与总结

（1）常见的C语言开发软件还有哪些？

（2）IAR Embedded Workbench提供了强大的配置功能，在下载程序的时候需要根据实际情况选择下载器型号以及下载方式。

项目四 智能鼠的基本功能操作

学习目标

（1）认识智能鼠人机交互系统。

（2）学习智能鼠的红外检测原理。

（3）学习智能鼠运动姿态的控制。

智能鼠的基本功能是从起点运行到终点。在宽度有限、大量转弯的迷宫当中快速而准确运行，离不开高精度的传感器检测以及电动机运行控制。不同的环境，光照强度不同，地面摩擦也有一定的差异，所以必须使用人机交互的手段，调试智能鼠的红外检测精度和电动机转速。

任务一 智能鼠的人机交互系统

一个好的人机交互系统应该把人脑的决策很快地传递给机器人，同时也应该很快地把系统信息反馈回来，以便使用者做出决策。在TQD-Micromouse-JD智能鼠的设计中，人机交互系统主要指的是调试用显示器的电路设计，其性能的优良直接影响智能鼠的驱动速度和运行状态，从而影响智能鼠走迷宫竞赛的成绩。

智能鼠的显示系统（LED显示电路）主要指的是输出电路中的键盘显示电路，可以显示智能鼠所处的迷宫坐标和采集的挡板信息；键盘还可以设置为单步验证各个功能模块，并在数码管上进行显示，比如步进电动机的转速和方向等。键盘显示电路原理图如图1-4-1所示。

视频

实验：数码管显示指定编号

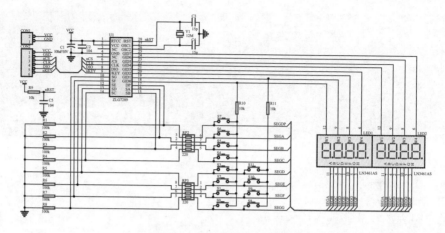

图1-4-1 键盘显示电路原理图

核心函数：Download_7289

```
/***************************************************************
** Function name:Download_7289
** Descriptions:下载数据
** input parameters: mode=0 (下载数据且按方式0译码)
**                   mode=1 (下载数据且按方式1译码)
**                   mode=2 (下载数据但不译码)
**                   number (数码管编号，取值范围0~7)
**                   dp=0 (小数点不亮)
**                   dp=1 (小数点亮)
**                   data (要显示的数据)
** output parameters:无
** Returned value:无
***************************************************************/
void Download_7289 (uchar mode, char number, char dp, char data)
{
    uchar modeDat[3]={0x80,0xC8,0x90};
    uchar temp_mode;
    uchar temp_data;

    if (mode>2) {
        mode=2;
    }

    temp_mode=modeDat[mode];
    number &=0x07;
    temp_mode|=number;
    temp_data=data&0x7F;
    if (dp==1) {
        temp_data|=0x80;
```

```
    }
    CmdDat_7289(temp_mode, temp_data);
}
```

流程图：整体思路简化为系统初始化和数码管显示数据两部分，如图1-4-2所示。

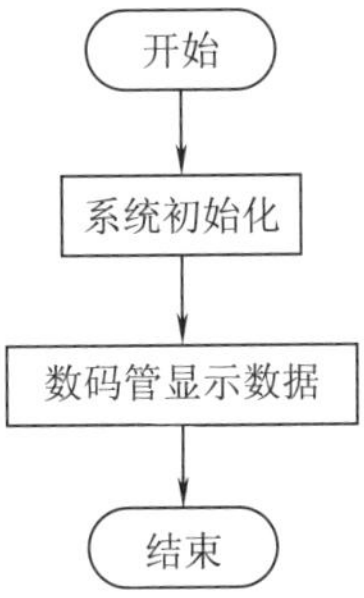

```
      ┌──────┐
      │ 开始 │
      └──┬───┘
         ↓
    ┌─────────┐
    │系统初始化│
    └────┬────┘
         ↓
   ┌──────────┐
   │数码管显示数据│
   └─────┬────┘
         ↓
      ┌──────┐
      │ 结束 │
      └──────┘
```

图1-4-2 数码管显示流程图

主程序：

```
                          main.c
/************************************************************
** Function name:main
** Descriptions:主函数
** input parameters:无
** output parameters:无
** Returned value:无
************************************************************/
main (void)
{
  SysCtlClockSet(SYSCTL_SYSDIV_4|SYSCTL_USE_PLL|SYSCTL_
     OSC_MAIN|SYSCTL_XTAL_6MHZ );  /*使能PLL,设定系统时钟频率为50MHz*/
     Init_7289();                  /*初始化ZLG7289*/
  /* 显示数据 */
  Download_7289(0, 0, 0, 1);  //8个数码管全部显示1
  Download_7289(0, 1, 0, 1);
  Download_7289(0, 2, 0, 1);
  Download_7289(0, 3, 0, 1);
  Download_7289(0, 4, 0, 1);
  Download_7289(0, 5, 0, 1);
  Download_7289(0, 6, 0, 1);
  Download_7289(0, 7, 0, 1);
while (1);
}
```

任务二　智能鼠的红外检测

传感器在控制系统中起了非常重要的作用,是感知系统中必不可少的部件。TQD-Micromouse-JD上共有五组红外传感器,每组红外传感器由红外线发射头和红外线接收头组成。

一、红外检测

智能鼠红外检测电路用于迷宫挡板的检测,分为左方、左斜、前方、右斜、右方五个方向,其具体作用如下:

（1）利用五组传感器检测一定范围内的障碍物,既可以判断一定距离的范围内是否存在障碍物,也可用于智能鼠运行过程中的迷宫格信息识别和转弯控制。

（2）左右两侧的红外传感器能够粗略判断障碍物的远近距离,可以指示出无障碍物、有障碍物和障碍物太近三种状态。

五个方向的传感器电路原理相同,其中一个方向的检测电路如图1-4-3所示。

RF2为红外发射头,W2为限流可调电阻,用来调节发射红外线的强度。TQD-Micromouse-JD使用的红外接收头型号为IRM8601S（见图1-4-4）,该接收头对载波频率为38 kHz的红外线信号最为敏感,探测距离也最远,IRM8601S的调制信号为周期1 200 μs的方波（见图1-4-5）,当它检测到有效红外线信号时输出低电平,否则输出高电平,如图1-4-6所示。

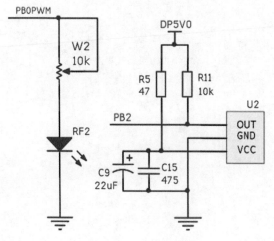

图1-4-3　红外检测电路

图1-4-4　IRM8601S红外接收头

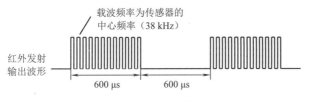

图1-4-5 驱动红外发射的调制波

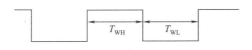

图1-4-6 传感器输出波形

视 频

实验：红外线传感器测距原理

如何控制红外线的发射强度呢？最直接的方法是改变驱动电流或者电压，调节电阻W2就可以改变驱动电流。还有没有其他方法呢？

前面曾提过，一体化接收头含有一个中心频率为38 kHz的带通滤波器，如图1-4-7所示，当红外线的载波频率为38 kHz时，经过滤波器衰减最小。越是偏离，衰减越多，这也是一体化的接收头抗干扰的关键原理。这里通过调节限流电阻和载波频率相结合的方法来调节红外传感器的检测距离。

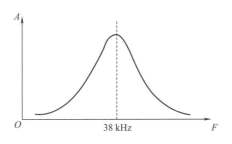

图1-4-7 带通滤波示意图

红外发射头可以发出红外线，所有的物体都可以不同程度地反射红外线。若距离合适，在经过迷宫挡板反射后可以被接收头接收。

同样可以使用中断函数，来读取红外传感器是否检测到挡板信息。TQD-Micromouse-JD智能鼠共有两组红外PWM发生器模块，分别为驱动两个45°红外发射的1号模块，以及驱动左方、前方、右方三个红外发射的2号模块。

```
                核心函数1：__irSendFreq
/*****************************************************
** Function name:__irSendFreq
** Descriptions:发射红外线
** input parameters:__uiFreq（红外线调制频率）
**                  __cNumber（选择需要设置的PWM模块）
** output parameters:无
```

```
** Returned value:无
***************************************************************/
void __irSendFreq (uint __uiFreq, char __cNumber)
{
    __uiFreq=SysCtlClockGet()/__uiFreq;
    switch(__cNumber){

    case 1:
        PWMGenPeriodSet(PWM_BASE, PWM_GEN_1, __uiFreq);
                                    /*设置PWM发生器1的周期*/
        PWMPulseWidthSet(PWM_BASE, PWM_OUT_2, __uiFreq/2);
                                    /*设置PWM2输出的脉冲宽度*/
        PWMGenEnable(PWM_BASE, PWM_GEN_1);
                                    /*使能PWM发生器1*/
        break;

    case 2:
        PWMGenPeriodSet(PWM_BASE, PWM_GEN_2, __uiFreq);
                                /*设置PWM发生器2的周期*/
        PWMPulseWidthSet(PWM_BASE, PWM_OUT_4, __uiFreq/2);
                                /*设置PWM4输出的脉冲宽度*/
        PWMGenEnable(PWM_BASE, PWM_GEN_2); /*使能PWM发生器2*/
        break;
    default:
        break;
    }
}
```

核心函数2：__irCheck

```
/***************************************************************
** Function name:__irCheck
** Descriptions:红外线传感器检测
** input parameters:无
** output parameters:无
** Returned value:无
***************************************************************/
void __irCheck (void)
{
    static uchar ucState=0;
    static uchar ucIRCheck;
    switch (ucState) {
    case 0:
        __irSendFreq(32200, 2);     /*检测左右两侧近距*/
        __irSendFreq(35000, 1);     /*驱动斜角上的传感器检测*/
        break;
    case 1:
```

```c
    ucIRCheck=GPIOPinRead(GPIO_PORTB_BASE, 0x3e);
                            /*读取传感器状态*/
    PWMGenDisable(PWM_BASE, PWM_GEN_2);
                            /*禁止PWM发生器2*/
    PWMGenDisable(PWM_BASE, PWM_GEN_1);
                            /*禁止PWM发生器1*/
    if (ucIRCheck&__RIGHTSIDE) {
        __GucDistance[__RIGHT]&=0xfd;
    } else {
        __GucDistance[__RIGHT]|=0x02;
    }
    if (ucIRCheck&__LEFTSIDE) {
        __GucDistance[__LEFT]&=0xfd;
    } else {
        __GucDistance[__LEFT]|=0x02;
    }
    if (ucIRCheck&__FRONTSIDE_R) {
        __GucDistance[__FRONTR]=0x00;
    } else {
        __GucDistance[__FRONTR]=0x01;
    }
    if (ucIRCheck&__FRONTSIDE_L) {
        __GucDistance[__FRONTL]=0x00;
    } else {
        __GucDistance[__FRONTL]=0x01;
    }
    break;
case 2:
    __irSendFreq(36000, 2);
                            /*驱动检测左前右三个方向远距*/
    break;
case 3:
    ucIRCheck=GPIOPinRead(GPIO_PORTB_BASE, 0x2a);
                            /*读取传感器状态*/
    PWMGenDisable(PWM_BASE, PWM_GEN_2);
                            /*禁止PWM发生器2*/
    break;
case 4:
    __irSendFreq(36000, 2);
                            /*重复检测左前右三个方向远距*/
    break;
case 5:
    ucIRCheck&=GPIOPinRead(GPIO_PORTB_BASE, 0x2a);
                            /*读取传感器状态*/
    PWMGenDisable(PWM_BASE, PWM_GEN_2);
                            /*禁止PWM发生器2*/
```

```
            if (ucIRCheck&__RIGHTSIDE) {
                __GucDistance[__RIGHT]&=0xfe;
            } else {
                __GucDistance[__RIGHT]|=0x01;
            }
            if (ucIRCheck&__LEFTSIDE) {
                __GucDistance[__LEFT]&=0xfe;
            } else {
                __GucDistance[__LEFT]|=0x01;
            }
            if (ucIRCheck&__FRONTSIDE) {
                __GucDistance[__FRONT]&=0xfe;
            } else {
                __GucDistance[__FRONT]|=0x01;
            }
            break;
        default:
            break;
    }
    ucState=(ucState+1)%6;              /*循环检测*/
}
```

流程图：根据传感器特性可知，红外发射频率越接近38 kHz，检测距离越远。所以使用两组不同的频率来驱动发射头，如32.2 kHz和36 kHz。红外测距流程图如图1-4-8所示。

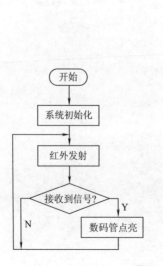
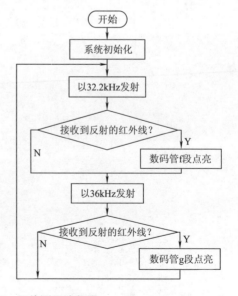

图1-4-8　红外测距流程图

主程序：

```
                        main.c
main (void)
{
  SysCtlClockSet(SYSCTL_SYSDIV_4|SYSCTL_USE_PLL|SYSCTL_OSC_MAIN|SYSCTL_
XTAL_6MHZ);
  /*使能PLL，设定系统时钟频率为50 MHz*/
    IRInit();                  /*传感器初始化*/
    SysTimerInit();            /*系统时钟初始化*/
    Init_7289();               /*7289初始化*/
    while(1);                  /*等待中断*/
}
```

二、红外调试步骤

（1）45°传感器调节。将智能鼠放在迷宫通道中间，调节第二个和第四个电位器，直到第二个和第四个数码管g段刚开始闪烁为止。

（2）左90°传感器调节。智能鼠紧贴右侧挡板，调节第一个电位器，直到第一个数码管g段完全点亮、f段较强闪烁为止。

（3）右90°传感器调节。智能鼠紧贴左侧挡板，调节第五个电位器，直到第五个数码管g段完全点亮、f段较强闪烁为止。

（4）正前方传感器调节。将智能鼠放在两个单元格交界处，调节第三个电位器，直到第三个数码管g段刚开始闪烁为止，如图1-4-9所示。

图1-4-9　45°传感器调节、左90°传感器调节、右90°传感器调节、正前方传感器调节

任务三　智能鼠的运动姿态控制

TQD-Micromouse-JD 智能鼠电动机驱动电路如图1-4-10所示。其中BA6845FS是步进电动机驱动芯片，包含两个H桥电路，最大驱动电流为1 A，且在输入逻辑的控制下输出有三种模式：正向、反向和停止。

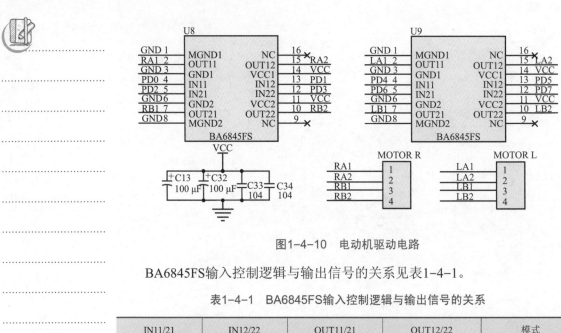

图1-4-10 电动机驱动电路

BA6845FS输入控制逻辑与输出信号的关系见表1-4-1。

表1-4-1 BA6845FS输入控制逻辑与输出信号的关系

IN11/21	IN12/22	OUT11/21	OUT12/22	模式
L	H	H	L	正向
H	H	L	H	反向
L	L	开路	开路	停止
H	L	开路	开路	停止

实验一 智能鼠跑起来

读者经过电动机驱动的学习,已经知道了电动机有多种参数需要考虑:

(1)电动机状态:启动还是停止。

(2)运行方向:向前还是向后。

(3)速度:快还是慢。

(4)需要转动的步数。

(5)已经转动的步数。

因此,可以建立一个函数结构体来存储这些参数。

结构体和其他基础数据类型一样(如int类型、char类型),只不过结构体可以根据需要进行自定义。结构体在函数中的作用是封装。封装的好处就是可以再次利用,让使用者不必关心结构体具体是什么,只要根据定义使用即可。

核心定义：电动机驱动

```
/***************************************************
    常量宏定义——电动机状态
****************************************************/
#define   MOTORSTOP          0         /*电动机停止*/
#define   WAITONESTEP        1         /*电动机暂停一步*/
#define   MOTORRUN           2         /*电动机运行*/
/***************************************************
    常量宏定义——电动机运行方向
****************************************************/
#define   MOTORGOAHEAD       0         /*电动机前进*/
#define   MOTORGOBACK        1         /*电动机后退*/
/***************************************************
    结构体定义
****************************************************/
struct  motor {
    char    cState;                    /*电动机运行状态*/
    char    cDir;                      /*电动机运行方向*/
    uint    uiPulse;                   /*电动机需要运行的脉冲*/
    uint    uiPulseCtr;                /*电动机已运行的脉冲*/
    int     iSpeed;                    /*当前速度*/
};
typedef struct  motor  MOTOR;
```

流程图：本实验根据智能鼠运行状态、运行方向、速度、需要转动的步数、已经转动的步数来控制智能鼠的运行，如图1-4-11所示。

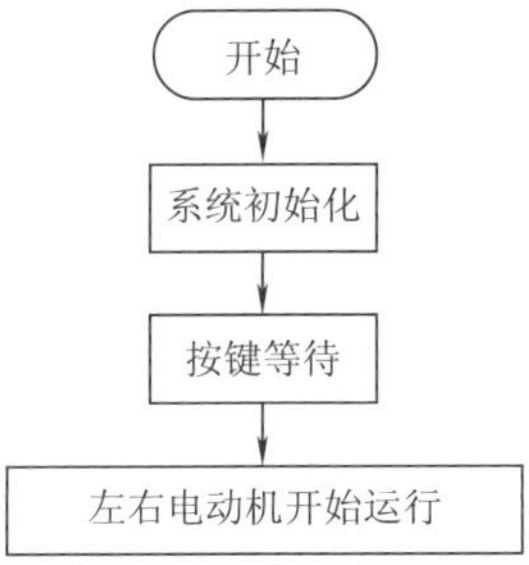

图1-4-11　智能鼠运行流程图

主程序：

```
                          main.c
/***************************************************
** Function name:main
** Descriptions:主函数
** input parameters:无
```

```
** output parameters:无
** Returned value:无
*************************************************************/
main (void)
{
    SysCtlClockSet(SYSCTL_SYSDIV_4|SYSCTL_USE_PLL|SYSCTL_OSC_
    MAIN|SYSCTL_XTAL_6MHZ);        /*使能PLL,设定系统时钟频率为50MHz*/

    SysCtlPeripheralEnable(SYSCTL_PERIPH_GPIOC);
                                    /*使能GPIO C口外设*/
    SysCtlPeripheralEnable(SYSCTL_PERIPH_GPIOD);
                                    /*使能GPIO D口外设*/

    __GmRight.iSpeed=SysCtlClockGet()/300;
                                    /*设定步进电动机每秒转动300步*/
    __GmLeft.iSpeed = SysCtlClockGet()/300;
                                    /*设定步进电动机每秒转动300步*/
    GPIODirModeSet(GPIO_PORTC_BASE, KEY, GPIO_DIR_MODE_IN);
                                    /*设置按键口为输入*/
    SysTickInit();         /*系统时钟初始化*/
    MotorInit();           /*传感器初始化*/

    while(1){
        if (KeyCheck()==true) {    /*查询判断是否有按键按下*/
    __GmRight.uiPulse=2000;            /*设定步进电动机转动2000步*/
    __GmRight.cDir=FORWARD;            /*设定步进电动机向前转动*/
    __GmRight.cState=RUN;              /*启动步进电动机转动*/
    __GmLeft.uiPulse=2000;             /*设定步进电动机转动2000步*/
    __GmLeft.cDir=FORWARD;             /*设定步进电动机向前转动*/
    __GmLeft.cState=RUN;               /*启动步进电动机转动*/
            while(__GmLeft.cState!=STOP&&__GmLeft.cState!=STOP);
        }
    }
}
```

视频
实验:电动机差速运行

实验二 电动机差速运行控制

智能鼠左右电动机速度相同时,它会沿着直线行走;否则,就会根据速度的差值大小转圈。下面就通过实验来尝试控制智能鼠转弯。

流程图:首先设定左右电动机的速度,根据速度差值来控制智能鼠转弯。

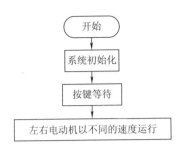

图1-4-12　智能鼠差速运行流程图

主程序：

```
                        main.c
/***************************************************************
** Function name:main
** Descriptions:主函数
** input parameters:无
** output parameters:无
** Returned value:无
***************************************************************/
main (void)
{

  SysCtlClockSet(SYSCTL_SYSDIV_4|SYSCTL_USE_PLL|SYSCTL_OSC_MAIN|
  SYSCTL_XTAL_6MHZ);              /*使能PLL,设定系统时钟频率为50MHz*/
    SysCtlPeripheralEnable(SYSCTL_PERIPH_GPIOC);
               /*使能GPIO C口外设*/
    SysCtlPeripheralEnable(SYSCTL_PERIPH_GPIOD);
               /*使能GPIO D口外设*/
    GPIODirModeSet(GPIO_PORTC_BASE, KEY, GPIO_DIR_MODE_IN);
               /*设置按键口为输入*/
    SysTickInit();     /*系统时钟初始化*/
    MotorInit();       /*步进电动机初始化*/
    while(1)
    {
if (KeyCheck()==true)              /*判断是否有按键按下*/
{
 __GmRight.iSpeed=SysCtlClockGet()/200;  /*设定步进电动机每秒转动200步*/
 __GmLeft.iSpeed=SysCtlClockGet()/100;   /*设定步进电动机每秒转动100步*/
 __GmRight.uiPulse=1000;                 /*设定步进电动机转动1000步*/
 __GmRight.cDir=FORWARD;                 /*设定步进电动机向前转动*/
 __GmRight.cState=RUN;                   /*启动步进电动机转动*/
 __GmLeft.uiPulse=500;                   /*设定步进电动机转动500步*/
 __GmLeft.cDir=FORWARD;                  /*设定步进电动机向前转动*/
 __GmLeft.cState=RUN;                    /*启动步进电动机转动*/
            TimerLoadSet(TIMER0_BASE, TIMER_A,__GmRight.iSpeed);
```

```
                    TimerLoadSet(TIMER1_BASE, TIMER_A,__GmLeft.iSpeed);
                    while(__GmRight.cState!=STOP&&__GmLeft.cState!=STOP);
            }
        }
    }
```

实验三 电动机运行路程控制

通过上面两个实验已经学会了电动机的驱动方法和速度控制方法,那么如何确定智能鼠在迷宫中的位置呢?修改"智能鼠跑起来"程序中的__GmRight.uiPulse和__GmLeft.uiPulse均为500,观察智能鼠的运行距离;再次修改为750观察运行距离。通过实验得出结论,在轮毂不变的情况下,智能鼠运行500步可以走过四个单元格,运行750步可以走过六个单元格。智能鼠正是通过记录运行的步数,结合轮毂直径、每个单元格需要的步数,得出运行的单元格数量的。

TQD-Micromouse-JD在使用标准轮毂的情况下,运行一个单元格需要125步。

核心函数1:mouseGoahead

```
/***************************************************************
** Function name:mouseGoahead
** Descriptions:前进N格
** input parameters:iNblock(前进的格数)
** output parameters:无
** Returned value:无
***************************************************************/
void mouseGoahead (char  cNBlock)
{
    char cL=0, cR=0, cCoor=1;
    if (__GmLeft.cState)                    //判断电动机状态
    {
        cCoor=0;
    }
    /*
     *   设定运行任务
     */
    __GucMouseState=__GOAHEAD;
    __GiMaxSpeed=MAXSPEED;
    __GmRight.cDir=__MOTORGOAHEAD;
    __GmLeft.cDir=__MOTORGOAHEAD;
```

```c
    __GmRight.uiPulse=__GmRight.uiPulse+cNBlock*ONEBLOCK-6;
    __GmLeft.uiPulse=__GmLeft.uiPulse+cNBlock*ONEBLOCK-6;
    __GmRight.cState=__MOTORRUN;
    __GmLeft.cState=__MOTORRUN;

while (__GmLeft.cState!=__MOTORSTOP) {
    if (__GmLeft.uiPulseCtr>=ONEBLOCK) {
                            /*判断是否走完一格*/
        __GmLeft.uiPulse-=ONEBLOCK;
        __GmLeft.uiPulseCtr-=ONEBLOCK;
        if (cCoor)
        {
            cNBlock--;
            __mouseCoorUpdate(); /*更新坐标*/
        }
        else
        {
            cCoor=1;
        }
    }
    if (__GmRight.uiPulseCtr>=ONEBLOCK) {
                            /*判断是否走完一格*/
        __GmRight.uiPulse-=ONEBLOCK;
        __GmRight.uiPulseCtr-=ONEBLOCK;
    }
    if (__GucDistance[__FRONT]) {
                            /*前方有挡板,则跳出程序*/
        __GmRight.uiPulse=__GmRight.uiPulseCtr+70;
        __GmLeft.uiPulse=__GmLeft.uiPulseCtr+70;
        while (__GucDistance[__FRONT]) {
            if ((__GmLeft.uiPulseCtr+20)>__GmLeft.uiPulse) {
                goto End;
            }
        }
    }
    if (cNBlock<2){
        if (cL) {           /*是否允许检测左边*/
            if ((__GucDistance[__LEFT]&0x01)==0) {
                            /*左边有支路,跳出程序*/
                __GmRight.uiPulse=__GmRight.uiPulseCtr+74;
                __GmLeft.uiPulse=__GmLeft.uiPulseCtr+74;
                while ((__GucDistance[__LEFT]&0x01)==0) {
                    if ((__GmLeft.uiPulseCtr+20)>__GmLeft.uiPulse) {
                        goto End;
                    }
                }
```

```c
                    }
                } else {                /*左边有挡板时开始允许检测左边*/
                    if (__GucDistance[__LEFT]&0x01) {
                        cL = 1;
                    }
                }
                if (cR) {               /*是否允许检测右边*/
                    if ((__GucDistance[__RIGHT]&0x01)==0) {
                                        /*右边有支路,跳出程序*/
                        __GmRight.uiPulse=__GmRight.uiPulseCtr+74;
                        __GmLeft.uiPulse=__GmLeft.uiPulseCtr+74;
                        while ((__GucDistance[__RIGHT]&0x01)==0) {
                            if ((__GmLeft.uiPulseCtr+20)>__GmLeft.uiPulse) {
                                goto End;
                            }
                        }
                    } else {
                        if (__GucDistance[__RIGHT]&0x01) {
                                   /*右边有挡板时开始允许检测右边*/
                            cR=1;
                        }
                    }
                }
            }
        /*
         *  设定运行任务,让智能鼠走到支路的中心位置
         */
End:    __mouseCoorUpdate();            /*更新坐标*/
}
```

核心函数2：__mouseCoorUpdate

```c
/******************************************************************
** Function name:__mouseCoorUpdate
** Descriptions:根据运行步数更新单元格数值
** input parameters:无
** output parameters:无
** Returned value:无
******************************************************************/
void __mouseCoorUpdate (void)
{

  if(GucMouseDir==0)
  {
     GmcMouse.cY++;
  }
   Download_7289(1, 3, 0, GmcMouse.cY%10);
}
```

流程图：本实验通过记录运行的步数来计算当前的单元格数量，并通过7289显示出来，如图1-4-13所示。

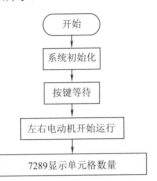

图1-4-13　智能鼠运行路程控制流程图

主程序：

```
                         main.c
/***************************************************************
** Function name:main
** Descriptions:主函数
** input parameters:无
** output parameters:无
** Returned value:无
***************************************************************/
main (void)
{
    uchar ucRoadStat=0;       /*统计某一坐标可前进的支路数*/
    mouseInit();              /*底层驱动的初始化*/
    Init_7289();              /*显示模块的初始化*/

    while (1) {
        switch (GucMouseTask) {    /*状态机处理*/
        case WAIT:
            sensorDebug();
            voltageDetect();
            delay(100000);
            if (keyCheck()==true) {  /*检测按键等待启动*/
                Reset_7289();        /*复位7289*/
                GucMouseTask=START;
            }
            break;
        case START:
         mouseGoahead(5);        //6、7、8
         while(1);
        default:
```

```
            break;
        }
    }
}
```

思考与总结

（1）智能鼠是如何实现人机交互的？

（2）红外传感器的作用是什么？如何提高检测精度？

（3）电动机共有哪几种类型？

（4）PWM技术常用作电动机的调速，具有响应迅速、精度高等优点。

第二篇 综合实践篇

本篇主要介绍智能鼠直线运动和精确转弯的调试方法，对智能鼠创新竞赛进行基本原理和操作的介绍，学生可以更快地掌握智能鼠的转弯函数、多种转弯组合的应用。参赛队员根据裁判随机抽取的任务试卷中的考核题目要求，现场编写程序实现对应功能。

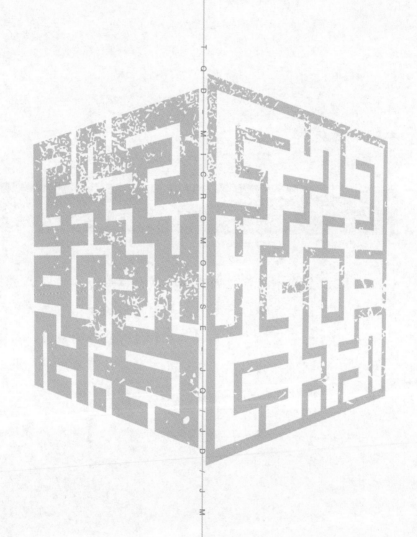

项目一

智能鼠高级功能

学习目标

(1) 掌握智能鼠直线运动的原理和控制方法。

(2) 学习控制智能鼠精确转弯运动。

智能鼠在迷宫中的运行,可以简化为直行和转弯两部分。直行是指智能鼠在迷宫中通过检测两侧挡板,校正车姿,避免碰触;转弯是指智能鼠精确转弯180°和90°,结合直行最终到达目的地。

任务一 智能鼠直行的控制方法

智能鼠上的传感器相当于它的"眼睛",而它的两个轮子相当于"脚"。想要它按照要求去运动和工作,就要学会控制使用它的"眼睛"和"脚"。人在运动时,通过眼睛观察周围的信息,并以这些信息作为依据去指挥脚运动。智能鼠也是这样,通过传感器来获取外界信息,并以此为依据来控制它的轮子运动。

当智能鼠向左偏移时,增加左侧电动机速度或者减小右侧电动机速度。

当智能鼠向右偏移时,增加右侧电动机速度或者减小左侧电动机速度。

这里采用"暂停一步"的方法:当智能鼠偏移时,外侧电动机暂停一步,从而达到校正的目的。

```
              核心函数1: Timer0A_ISR
/****************************************************************
** Function name:Timer0A_ISR
** Descriptions:Timer0中断服务函数
** input parameters:无
** output parameters:无
** Returned value:无
****************************************************************/
```

视频

实验:智能鼠直行

```c
void Timer0A_ISR(void)
{
    static char n=0,m=0;
    TimerIntClear(TIMER0_BASE, TIMER_TIMA_TIMEOUT);
                            /*清除定时器0中断*/
    switch (__GmRight.cState) {
    case __MOTORSTOP:            /*停止,同时清零速度和脉冲值*/
        __GmRight.iSpeed=0;
        __GmRight.uiPulse=0;
        __GmRight.uiPulseCtr=0;
        break;
    case __WAITONESTEP:          /*暂停一步*/
        __GmRight.cState=__MOTORRUN;
        break;
    case __MOTORRUN:     /*电动机运行*/
        if(__GucMouseState==__GOAHEAD) {
                            /*根据传感器状态微调电动机位置*/
            if(__GucDistance[__FRONTL]&&(__GucDistance[__FRONT]==0)){
                if (n==1) {
                    __GmRight.cState=__WAITONESTEP;
                }
                n++;
                n%=2;   /*运行一步,暂停一步,相当于速度减半*/
            } else {
                n=0;
            }

            if((__GucDistance[__RIGHT]==1)&&(__GucDistance[__LEFT]==0))
            {
                if(m==1)
                {
                    __GmRight.cState=__WAITONESTEP;
                }
                m++;
                m%=2;
            } else
            {
                m=0;
            }
        }
        __rightMotorContr();  /*电动机驱动程序*/
        break;

    default:
```

```
            break;
    }
    /*
     *  是否完成任务判断
     */
    if (__GmRight.cState!=__MOTORSTOP) {
        __GmRight.uiPulseCtr++;        /*运行脉冲计数*/
        __speedContrR();               /*速度调节*/
        if (__GmRight.uiPulseCtr>=__GmRight.uiPulse) {
            __GmRight.cState=__MOTORSTOP;
            __GmRight.uiPulseCtr=0;
            __GmRight.uiPulse=0;
            __GmRight.iSpeed=0;
        }
    }
}
                    核心函数2：Timer1A_ISR
/***************************************************************
** Function name:Timer1A_ISR
** Descriptions:Timer1中断服务函数
** input parameters:__GmLeft.cState（驱动步进电动机的时序状态）
**                  __GmLeft.cDir（步进电动机运动的方向）
** output parameters:无
** Returned value:无
***************************************************************/
void Timer1A_ISR(void)
{
    static char n=0, m=0;
    TimerIntClear(TIMER1_BASE, TIMER_TIMA_TIMEOUT);
                                /*清除定时器1中断*/
    switch (__GmLeft.cState) {
    case __MOTORSTOP:           /*停止，同时清零速度和脉冲值*/
        __GmLeft.iSpeed=0;
        __GmLeft.uiPulse=0;
        __GmLeft.uiPulseCtr=0;
        break;
    case __WAITONESTEP:         /*暂停一步*/
        __GmLeft.cState=__MOTORRUN;
        break;
    case __MOTORRUN:            /*电动机运行*/
        if (__GucMouseState==__GOAHEAD) {
                                /*根据传感器状态微调电动机位置*/
      if(__GucDistance[__FRONTR]&&(__GucDistance[__FRONT]==0)){
                if (n==1) {
```

```
                    __GmLeft.cState=__WAITONESTEP;
                }
                n++;
                n%=2;
            } else {
                n=0;
            }
    if ((__GucDistance[__LEFT]==1)&&(__GucDistance[__RIGHT]==0)){
                if(m==1) {
                    __GmLeft.cState=__WAITONESTEP;
                }
                m++;
                m%=2;
            } else {
                m=0;
            }
        }
        __leftMotorContr();          /*电动机驱动程序*/
        break;

    default:
        break;
    }
    /*
     * 是否完成任务判断
     */
    if (__GmLeft.cState!=__MOTORSTOP) {
        __GmLeft.uiPulseCtr++;       /*运行脉冲计数*/
        __speedContrL();             /*速度调节*/
        if (__GmLeft.uiPulseCtr>=__GmLeft.uiPulse) {
            __GmLeft.cState=__MOTORSTOP;
            __GmLeft.uiPulseCtr=0;
            __GmLeft.uiPulse=0;
            __GmLeft.iSpeed=0;
        }
    }
}
```

流程图如图2-1-1所示。

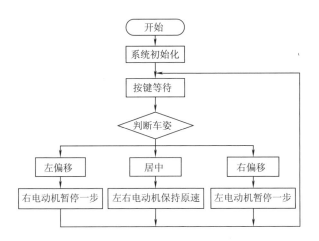

图2-1-1 智能鼠运行校正流程图

主程序：

```
                        main.c
/***************************************************************
** Function name:main
** Descriptions:主函数
** input parameters:无
** output parameters:无
** Returned value:无
***************************************************************/
main (void)
{
    uchar n=0;                  /*有多个支路的坐标的数量*/
    uchar ucRoadStat=0;         /*统计某一坐标可前进的支路数*/
    uchar ucTemp=0;             /*用于START状态中坐标转换*/
    mouseInit();                /*底层驱动的初始化*/
    Init_7289();                /*显示模块的初始化*/
    while (1) {
        switch (GucMouseTask) {  /*状态机处理*/
        case WAIT:
            sensorDebug();
            voltageDetect();
            delay(100000);
            if (keyCheck()==true) {   /*检测按键等待启动*/
                Reset_7289();          /*复位ZLG7289*/
                GucMouseTask=START;
            }
            break;
        case START:
            crosswayChoice();  /*用右手法则搜索选择前进方向*/
```

```
                mazeSearch();        /*前进一格*/
            break;
        default:
            break;
        }
    }
}
```

任务二 智能鼠精确转弯的控制方法

智能鼠在竞赛迷宫中运行，除直行外还会遇到大量的转弯，例如，90°、180°转弯。快速而准确地转弯是智能鼠提高成绩的有效方法。

核心函数1：mouseTurnright

```
/***************************************************************
** Function name:mouseTurnright
** Descriptions:右转
** input parameters:无
** output parameters:无
** Returned value:无
***************************************************************/
void mouseTurnright(void)
{
    while (__GmLeft.cState!=__MOTORSTOP);
    while (__GmRight.cState!=__MOTORSTOP);
    /*开始右转*/
    __GucMouseState=__TURNRIGHT;
    __GmRight.cDir=__MOTORGOBACK;
    __GmRight.uiPulse=45;
    __GmLeft.cDir=__MOTORGOAHEAD;
    __GmLeft.uiPulse=45;
    __GmRight.cState=__MOTORRUN;
    __GmLeft.cState=__MOTORRUN;
GucMouseDir=(GucMouseDir+1)%4;
    /*方向标记*/
    while (__GmLeft.cState!=__MOTORSTOP);
    while (__GmRight.cState!=__MOTORSTOP);
    __mazeInfDebug();
    __delay(100000);
}
```

核心函数2：mouseTurnleft

```
/***************************************************************
** Function name:mouseTurnleft
```

```
**  Descriptions:左转
**  input parameters:无
**  output parameters:无
**  Returned value:无
*************************************************************/
void mouseTurnleft(void)
{
    while (__GmLeft.cState!=__MOTORSTOP);
    while (__GmRight.cState!=__MOTORSTOP);
    /*开始左转*/
    __GucMouseState=__TURNLEFT;
    __GmRight.cDir=__MOTORGOAHEAD;
    __GmRight.uiPulse=47;
    __GmLeft.cDir=__MOTORGOBACK;
    __GmLeft.uiPulse=47;
    __GmRight.cState=__MOTORRUN;
    __GmLeft.cState=__MOTORRUN;
    GucMouseDir=(GucMouseDir+3)%4;
    /*方向标记*/
    while (__GmLeft.cState!=__MOTORSTOP);
    while (__GmRight.cState!=__MOTORSTOP);
    __mazeInfDebug();
    __delay(100000);
}
```

核心函数3：mouseTurnback

```
/*************************************************************
**  Function name:mouseTurnback
**  Descriptions:后转
**  input parameters:无
**  output parameters:无
**  Returned value:无
*************************************************************/
void mouseTurnback(void)
{
    /*等待停止*/
    while (__GmLeft.cState!=__MOTORSTOP);
    while (__GmRight.cState!=__MOTORSTOP);
    /*开始后转*/
    __GucMouseState=__TURNBACK;
    __GmRight.cDir=__MOTORGOBACK;
    __GmRight.uiPulse=90;//162*10;
    __GmLeft.cDir=__MOTORGOAHEAD;
    __GmLeft.uiPulse=90;//162*10;
    __GmLeft.cState=__MOTORRUN;
    __GmRight.cState=__MOTORRUN;
```

```
    GucMouseDir=(GucMouseDir+2)%4;
    /*方向标记*/
    while(__GmLeft.cState!=__MOTORSTOP);
    while(__GmRight.cState!=__MOTORSTOP);
    __mazeInfDebug();
    __delay(100000);
}
```

流程图：以右转为例来实现智能鼠转弯角度的精确控制，如图2-1-2所示。

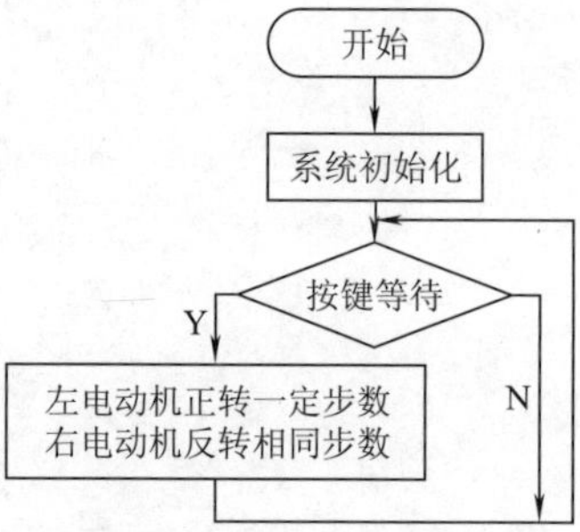

图2-1-2 智能鼠转弯角度精确控制

主程序：

```
                        main.c
/************************************************************
** Function name:main
** Descriptions:主函数
** input parameters:无
** output parameters:无
** Returned value:无
************************************************************/
main (void)
{
    uchar n=0;               /*有多个支路的坐标的数量*/
    uchar ucRoadStat=0;      /*统计某一坐标可前进的支路数*/
    uchar ucTemp=0;          /*用于START状态中坐标转换*/
    mouseInit();             /*底层驱动的初始化*/
    Init_7289();             /*显示模块的初始化*/
    while (1) {
        switch (GucMouseTask) {   /*状态机处理*/
        case WAIT:
            sensorDebug();
            voltageDetect();
```

```
            delay(100000);
            if (keyCheck()==true) {/*检测按键等待启动*/
                Reset_7289();           /*复位ZLG7289*/
                GucMouseTask=START;
            }
            break;
        case START:
            mouseTurnright();                   /*右转弯*/
            //mouseTurnback();
            //mouseTurnleft();
            break;
        default:
            break;
    }
}
```

思考与总结

（1）TQD-Micromouse-JD智能鼠如何实现转弯？

（2）逆时针转180°和顺时针转180°有何区别？

（3）为了避免相互之间的干扰，五组红外传感器采用交替发射的方式。高频率检测远距离障碍物，低频率检测近距离障碍物，多组数据共同判断障碍物信息，从而做到智能鼠速度和位置的高精度控制。

项目二 智能鼠实战任务

学习目标

（1）掌握智能鼠基本功能。
（2）学习智能鼠红外传感、中断函数、电动机驱动的运用。
（3）学习智能鼠转弯函数的使用和控制方法。
（4）学习智能鼠多种转弯组合应用的实现方法。

在智能鼠比赛中，为了考察参赛选手的真实水平，除了竞速环节外，还会设置现场编程项目。根据裁判现场给出的任务，编写程序并实现相应功能。

任务一 7289调试模块的使用

一、任务说明

本任务考查学生对于编程环境的熟悉程度，以及数码管显示函数的知识。编程并实现7289调试板显示指定数据，如图2-2-1所示。

图2-2-1 7289显示指定数据

二、任务目的

学习7289显示单元，利用7289显示板显示任意数据。

三、任务内容

编程并实现八个数码管依次显示0，1，2，…，7。

四、应用原理

在智能鼠的人机交互系统任务中，已经学习过7289软件包的应用，本任务尝试修改待显示的数据。

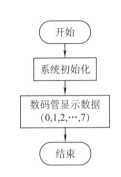

五、任务流程

实验的整体思路简化为系统初始化和数码管显示数据两部分，如图2-2-2所示。

图2-2-2 数码管显示指定数据流程图

六、具体步骤

步骤1：打开IAR EWARM集成开发环境，新建项目工程。

步骤2：按照软件使用方法，将对应章节目录文件夹下的头文件加入工程当中。

步骤3：将程序下载到"智能鼠"中。

步骤4：待下载完成后，断开智能鼠和计算机的连接，打开电源，运行程序，观察数码管的显示。

七、应用程序

```
                    程序清单 main.c
/*****************************************************
** Function name:main
** Descriptions:主函数
** input parameters:无
** output parameters:无
** Returned value:无
*****************************************************/
main (void)
{
   SysCtlClockSet(SYSCTL_SYSDIV_4|SYSCTL_USE_PLL|SYSCTL_OSC_
MAIN|SYSCTL_XTAL_6MHZ);          /*使能PLL，设定系统时钟频率为50MHz*/
    Init_7289();                 /*初始化ZLG7289*/
    /*显示数据*/
    Download_7289(0, 0, 0, 0);   /*第0号数码管显示0*/
    Download_7289(0, 1, 0, 1);   /*第1号数码管显示1*/
    Download_7289(0, 2, 0, 2);   /*第2号数码管显示2*/
```

```
        Download_7289(0, 3, 0, 3);        /*第3号数码管显示3*/
        Download_7289(0, 4, 0, 4);        /*第4号数码管显示4*/
        Download_7289(0, 5, 0, 5);        /*第5号数码管显示5*/
        Download_7289(0, 6, 0, 6);        /*第6号数码管显示6*/
        Download_7289(0, 7, 0, 7);        /*第7号数码管显示7*/
    while (1);
    }
```

任务二　非接触式启动和停止控制

一、任务说明

本任务考查学生对于红外函数、中断函数、电动机驱动知识的运用情况。

编程并实现非接触的方式（除了电源开关，不允许使用其他按键）启动智能鼠。智能鼠运行全程不得碰触任意挡板，在运行一段距离后，用同样非接触的方式停止运行。如图2-2-3所示的两个停车位置。

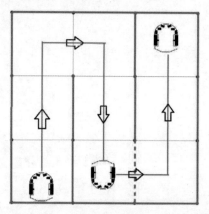

图2-2-3　非接触式启动和停止智能鼠

二、任务目的

学会使用非接触的方式启动和停止智能鼠。

三、任务内容

编程并实现，遮挡前方传感器，智能鼠启动；当三面有挡板时，智能鼠停车。

四、应用原理

红外检测障碍物信息可以作为智能鼠启动的依据，同样当三面有挡板时可以作为智能鼠停车的判断条件。

● 视　频

非接触式启动、运行和停止智能鼠

五、任务流程

非接触式启动和停止智能鼠流程图如图2-2-4所示。

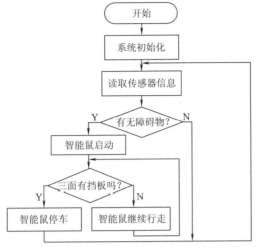

图2-2-4　非接触式启动和停止智能鼠流程图

六、具体步骤

步骤1：打开IAR EWARM集成开发环境，新建项目工程。

步骤2：按照软件使用方法，将对应章节目录文件夹下的头文件加入工程当中。

步骤3：将程序下载到"智能鼠"中。

步骤4：待下载完成后，断开智能鼠和计算机的连接，打开电源，运行程序。用手遮挡前方传感器，智能鼠启动；当三面有挡板时，智能鼠停车。

七、应用程序

```
              程序清单：修改后的mouseTurnback
/***********************************************************
** Function name:mouseTurnback
** Descriptions:后转
** input parameters:无
** output parameters:无
** Returned value:无
***********************************************************/
void mouseTurnback(void)
{
```

```
/*
 *等待停止
 */
while(__GmLeft.cState!=__MOTORSTOP);
while(__GmRight.cState!=__MOTORSTOP);
/*
 * 等待按键，否则一直处于停止状态
 */
while (1) {
        if (keyCheck()==true) {
                break;
        }
}
......
}
```

任务三　"8字型"路径运行控制

一、任务说明

本任务考查学生对多种转弯的组合应用。

编程并实现图2-2-5所示的智能鼠运行轨迹，类似数字8。智能鼠全程不得碰触任一挡板，最终回到出发位置。（启动时的方向，向左、向右均可。）

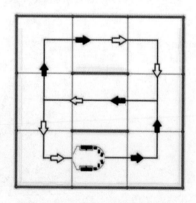

图2-2-5　智能鼠运行轨迹

二、任务目的

学习多种转弯的实现方法。

三、任务内容

编程并实现智能鼠按照8字轨迹运行。

四、应用原理

通过以上知识的学习,已经了解智能鼠在运行时可以依据自身算法进行转弯。分析图2-2-5,智能鼠在多路口时的转弯方向有左也有右,所以需要根据转弯的次数调用不同的转弯函数。

五、任务流程

本任务的整体思路简化为转弯函数的调用。只有一个路口时,不需要额外判断,所以不在考虑之内,如图2-2-6所示。

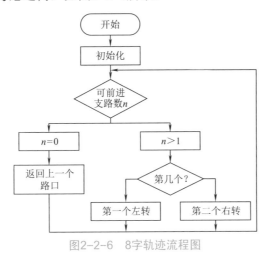

图2-2-6 8字轨迹流程图

六、具体步骤

步骤1:打开IAR EWARM集成开发环境,新建项目工程。

步骤2:按照软件使用方法,将对应章节目录文件夹下的头文件加入工程当中。

步骤3:将程序下载到"智能鼠"中。

步骤4:待下载完成后,断开智能鼠和计算机的连接,打开电源,运行程序。

七、应用程序

```
                程序清单 crosswayChoice
/**************************************************************
** Function name:crosswayChoice
** Descriptions:选择一条支路作为前进方向
```

```
***************************************************************/
void crosswayChoice (void)
{
  switch (SEARCHMETHOD) {    // 依据SEARCHMETHOD判断转弯方向
    case RIGHTMETHOD:
        mouseTurnright();              //右转弯
        break;
    case LEFTMETHOD:
        mouseTurnleft();               //左转弯
        break;
    case CENTRALMETHOD:
        centralMethod();
        break;
    case FRONTRIGHTMETHOD:
        frontRightMethod();
        break;
    default:
    break;
    }
}
```

思考与总结

（1）非接触式的启动和停止方法还有哪些种类？

（2）智能鼠采用两轮差速的转弯方法，速度差值越大，持续步数越大，转弯角度就越大；速度差值越小，持续步数越小，转弯角度就相应越小。

第三篇 拓展竞技篇

前两篇中已经介绍了智能鼠的软硬件以及智能鼠的基础编程调试方法。针对IEEE国际标准Micromouse走迷宫竞赛的要求，本篇主要介绍智能鼠优化算法。掌握智能鼠走迷宫竞赛的规范，可以最快的速度完成迷宫搜索和最优路径选择；分析智能鼠竞赛迷宫案例要点，为参加IEEE国际标准Micromouse走迷宫竞赛做好准备。

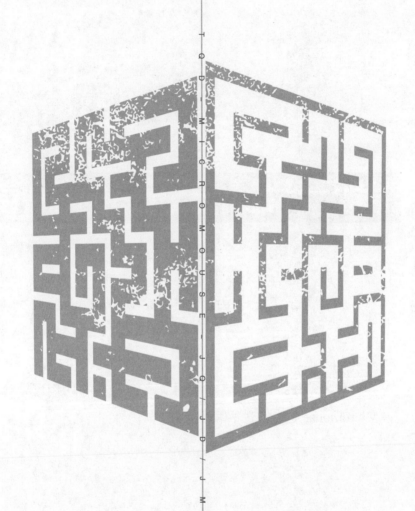

项目一

智能鼠的路径规划和行为决策算法

学习目标

（1）学习智能鼠的路径规划和决策算法。
（2）学习智能鼠路径规划的原理。

"智能鼠"如何才能在迷宫中快速运行呢？智能鼠的主要任务是根据IEEE国际标准Micromouse走迷宫竞赛规则完成迷宫搜索和最优路径的选择，是考察一个系统对一个未知环境的探测、分析及决策能力的一种比赛。下面来简单了解一下这方面的知识。

任务一　迷宫搜索的常用策略

迷宫搜索方法：在没有预知迷宫路径的情况下，智能鼠必须要先探索迷宫中的所有单元格，直到抵达终点为止。做这个处理的智能鼠要随时知道自己的位置及姿势，同时要记录下所有访问过的迷宫格四周是否有挡板。在搜索过程中为了节约搜索时间，还要尽量避免重复搜索。

那么，怎样来探索迷宫呢？通常有两种策略：（1）尽快到达终点；（2）搜索整个迷宫。

这两种策略各有利弊。利用第一种策略虽然可以缩短探索迷宫所需的时间，但是不一定能够得到整个迷宫地图的资料。若找到的路不是迷宫的最优路径，这将会影响智能鼠最后冲刺的时间。利用第二种策略，可以得到整个迷宫地图的资料，这样就可以求出最优路径。不过采用这种策略所使用的搜索时间较长。

任务二　迷宫搜索的基本法则

常用的搜索法则有三种：右手法则、左手法则和中心法则，如图3-1-1所示。

智能鼠搜索迷宫常用算法解析（右手）

智能鼠搜索迷宫常用算法解析（左手）

智能鼠搜索迷宫常用算法解析（中心）

（a）右手法则　　　　　　　（b）左手法则　　　　　　　（c）中心法则

图3-1-1　右手法则、左手法则、中心法则

右手法则：当智能鼠有多个可供选择的前进方向时，优先向右转，其次直行，最后左转；

左手法则：当智能鼠有多个可供选择的前进方向时，优先向左转，其次直行，最后右转；

中心法则：当智能鼠有多个可供选择的前进方向时，优先朝向终点的方向转动。

核心函数1：右手法则

```
/*****************************************************************
** Function name:rightMethod
** Descriptions:右手法则，优先向右前进
** input parameters:无
** output parameters:无
** Returned value:无
*****************************************************************/
void rightMethod (void)
{
    if ((GucMapBlock[GmcMouse.cX][GmcMouse.cY]&MOUSEWAY_R)&&
                            /*智能鼠的右侧有路口*/
        (mazeBlockDataGet(MOUSERIGHT)==0x00)) {
                            /*智能鼠的右边没有走过*/
        mouseTurnright();    /*智能鼠向右转*/
        return;
    }
    if ((GucMapBlock[GmcMouse.cX][GmcMouse.cY]&MOUSEWAY_F)&&
                            /*智能鼠的前方有路口*/
        (mazeBlockDataGet(MOUSEFRONT)==0x00)) {
                            /*智能鼠的前方没有走过*/
        return;              /*智能鼠不用转弯*/
    }
    if ((GucMapBlock[GmcMouse.cX][GmcMouse.cY]&MOUSEWAY_L)&&
                            /*智能鼠的左边有路口*/
        (mazeBlockDataGet(MOUSELEFT)==0x00)) {
```

```
                              /*智能鼠的左边没有走过*/
        mouseTurnleft();      /*智能鼠向左转*/
        return;
    }
}
```

核心函数2：左手法则

```
/***************************************************************
** Function name:leftMethod
** Descriptions:左手法则，优先向左运动
** input parameters:无
** output parameters:无
** Returned value:无
***************************************************************/
void leftMethod (void)
{
    if ((GucMapBlock[GmcMouse.cX][GmcMouse.cY]&MOUSEWAY_L)&&
                              /*智能鼠的左边有路口*/
        (mazeBlockDataGet(MOUSELEFT)==0x00)) {
                              /*智能鼠的左边没有走过*/
        mouseTurnleft();      /*智能鼠向左转*/
        return;
    }
    if ((GucMapBlock[GmcMouse.cX][GmcMouse.cY]&MOUSEWAY_F)&&
                              /*智能鼠的前方有路口*/
        (mazeBlockDataGet(MOUSEFRONT)==0x00)) {
                              /*智能鼠的前方没有走过*/
        return;               /*智能鼠不用转弯*/
    }
    if ((GucMapBlock[GmcMouse.cX][GmcMouse.cY]&MOUSEWAY_R)&&
                              /*智能鼠的右边有路口*/
        (mazeBlockDataGet(MOUSERIGHT)==0x00)) {
                              /*智能鼠的右边没有走过*/
        mouseTurnright();     /*智能鼠向右转*/
        return;
    }
}
```

核心函数3：中心法则

```
/***************************************************************
** Function name:centralMethod
** Descriptions:中心法则，根据智能鼠目前在迷宫中所处的位置决定使用何种
               搜索法则
** input parameters:无
** output parameters:无
** Returned value:无
***************************************************************/
void centralMethod (void)
{
```

```c
            if (GmcMouse.cX&0x08) {
                if (GmcMouse.cY&0x08) {
                    /*
                     * 此时智能鼠在迷宫的右上角
                     */
                    switch (GucMouseDir) {
                        case UP:                        /*当前智能鼠向上*/
                            leftMethod();               /*左手法则*/
                            break;
                        case RIGHT:                     /*当前智能鼠向右*/
                            rightMethod();              /*右手法则*/
                            break;
                        case DOWN:                      /*当前智能鼠向下*/
                            frontRightMethod();         /*中右法则*/
                            break;
                        case LEFT:                      /*当前智能鼠向左*/
                            frontLeftMethod();          /*中左法则*/
                            break;
                        default:
                            break;
                    }
                } else {
                    /*
                     * 此时智能鼠在迷宫的右下角
                     */
                    switch (GucMouseDir) {
                        case UP:                        /*当前智能鼠向上*/
                            frontLeftMethod();          /*中左法则*/
                            break;
                        case RIGHT:                     /*当前智能鼠向右*/
                            leftMethod();               /*左手法则*/
                            break;
                        case DOWN:                      /*当前智能鼠向下*/
                            rightMethod();              /*右手法则*/
                            break;
                        case LEFT:                      /*当前智能鼠向左*/
                            frontRightMethod();         /*中右法则*/
                            break;
                        default:
                            break;
                    }
                }
            } else {
                if (GmcMouse.cY&0x08) {
                    /*
                     * 此时智能鼠在迷宫的左上角
                     */
                    switch (GucMouseDir) {
                        case UP:                        /*当前智能鼠向上*/
```

```
            rightMethod();        /*右手法则*/
            break;
        case RIGHT:                /*当前智能鼠向右*/
            frontRightMethod();    /*中右法则*/
            break;
        case DOWN:                 /*当前智能鼠向下*/
            frontLeftMethod();     /*中左法则*/
            break;
        case LEFT:                 /*当前智能鼠向左*/
            leftMethod();          /*左手法则*/
            break;
        default:
            break;
        }
    } else {
        /*
         * 此时智能鼠在迷宫的左下角
         */
        switch (GucMouseDir) {
        case UP:                   /*当前智能鼠向上*/
            frontRightMethod();    /*中右法则*/
            break;
        case RIGHT:                /*当前智能鼠向右*/
            frontLeftMethod();     /*中左法则*/
            break;
        case DOWN:                 /*当前智能鼠向下*/
            leftMethod();          /*左手法则*/
            break;
        case LEFT:                 /*当前智能鼠向左*/
            rightMethod();         /*右手法则*/
            break;
        default:
            break;
        }
    }
}
```

思考与总结

（1）左手法则、右手法则和中心法则各有什么优缺点？

（2）智能鼠的搜索法则都是由左手和右手法则组合而成的。依据一定的规则，将迷宫分为若干部分，智能鼠位置不同、朝向不同，选用的转弯就不同。

项目二 智能鼠路径规划的原理

学习目标

（1）学习迷宫挡板信息的存储方法。

（2）学习等高图以及转弯加权的原理，尝试在竞赛迷宫图上手绘智能鼠最短路径。

任务一　迷宫信息的存储方法

进行路径规划首先需要记录所有位置的挡板信息。很明显，建立二维数组对整个迷宫进行坐标定义是非常有效的一个方法。每个单元格定义为一个坐标，该坐标对应的挡板信息均存储在建立的二维数组中。

当智能鼠到达一个单元格坐标时，应根据传感器检测结果记录下当前方格的挡板资料，为了方便管理和节省存储空间，每一个字节变量的低四位分别用来存储一个方格四周的挡板资料，迷宫共有16×16个方格，所以可以定义一个16×16的二维数组变量来保存整个迷宫挡板资料，如图3-2-1所示。

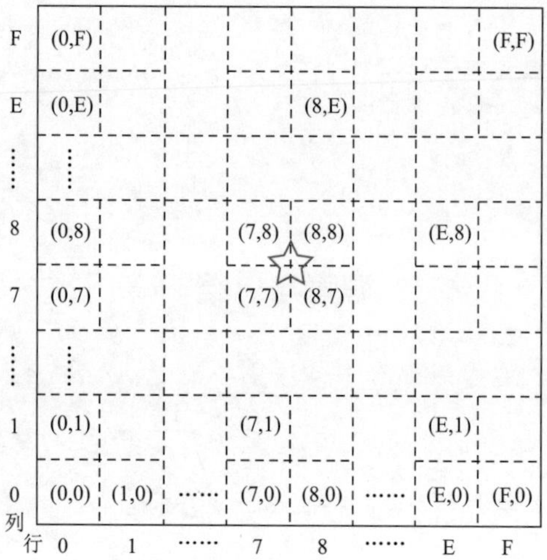

图3-2-1　迷宫坐标定义

首先将迷宫挡板资料全部初始化为0。凡是走过的迷宫格至少有一方没有挡板，也就是墙壁资料不全为0，这样就可以通过单元格存储的挡板资料是否为0来确定该单元格是否曾搜索过。挡板资料存储方式见表3-2-1。

表3-2-1 挡板资料存储方式

变量	方位	是否有挡板
bit0	上方0	1：无挡板，0：有挡板
bit1	右方1	1：无挡板，0：有挡板
bit2	下方2	1：无挡板，0：有挡板
bit3	左方3	1：无挡板，0：有挡板
bit7~bit4		保留位

任务二 等高图的制作方法

假设智能鼠已经搜索完整个迷宫或者只搜索了包含起点和终点的部分迷宫，且记录了已走过的每个迷宫格的挡板资料，那么它怎样根据已有信息找出一条从起点到终点最优的路径呢？下面引入等高图的概念和制作方法。

等高图就是等高线地图的简称，有如一般地图可以标出同一高度的地区范围，或有如气象报告时的等气压图，可以标出相等气压的范围及大小。

那么等高图运用在迷宫地图上，可以计算每一个迷宫格与迷宫终点的距离值，持续到计算出迷宫起点与迷宫终点的距离值为止。根据每个迷宫格与迷宫终点的距离值，再按由大到小的方式排列，这样就能在迷宫中找出一条最短路径。

起点标记为1，根据每个单元格的挡板信息，在该单元格上标识出距离起点的最短步数，从而得到任意坐标到起点的最优路径，如图3-2-2所示。

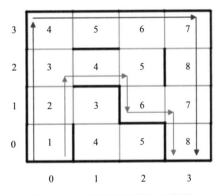

图3-2-2 等高图最终的示意图

思考与总结

（1）智能鼠是如何记录各单元格挡板信息的？

（2）智能鼠如何规划最优路径？

（3）智能鼠在转弯时需要进行减速和加速，所以需要对转弯进行加权。直行优先于转弯，长直道优先于短直道。

视频
智能鼠寻找最优路径的方法—等高图

项目三 智能鼠走迷宫程序设计

学习目标

学习智能鼠走迷宫的程序设计。

TQD-Micromouse-JD智能鼠的灵活性和智能程度不仅取决于硬件的结构和性能，还取决于程序设计的完整性。越是智能，其程序设计就越复杂。在智能鼠程序设计中，可以把程序结构简单分为两层，即底层驱动程序和顶层算法程序。

底层驱动程序主要实现智能鼠的一些基本功能，比如控制其直线前进N个单位坐标格，测量它前进的距离，向右或者向左转90°、向右转、防止碰撞挡板、迷宫格四周挡板信息的检测等。

顶层算法程序则主要是一些智能鼠的算法，如根据迷宫信息决定智能鼠动作，记住所走过迷宫的位置，寻找到达目的地的最优路径等。

任务一　运行姿态程序控制

智能鼠在运行后，会进行大量信息交换并切换状态。

1）等待状态

在该状态中，智能鼠静止在起点，等待开始命令。同时，实时显示传感器检测结果和电池的电压，这样方便调试传感器灵敏度和更换电池。当控制启动的按键按下后，智能鼠进入启动状态。

2）启动状态

在该状态中，智能鼠根据第一次转弯的方向判断起点是在坐标的（0，0）点还是（F，0）点。判断起点坐标程序流程图如图3-3-1所示。

3）搜索迷宫状态

在该状态中，智能鼠的任务就是探索并记忆迷宫地图。这里采用右手法则，并搜索全迷宫。迷宫搜索流程图如图3-3-2所示。

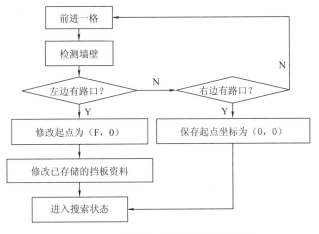

图3-3-1 判断起点坐标程序流程图

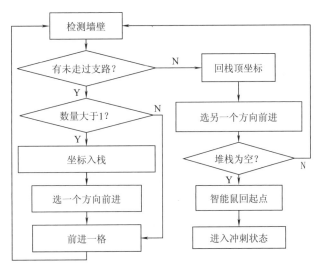

图3-3-2 迷宫搜索流程图

4）冲刺状态

迷宫搜索完毕后，根据算法找出一条最优路径冲刺到终点。冲刺结束后返回到起点。

任务二 基本程序结构解析

下面为读者介绍智能鼠运行时所调用的几个主要函数。

1）迷宫搜索程序

智能鼠每到一个坐标都会运行mazeSearch函数，检测四周的挡板信息并存储下来。

核心函数1: mazeSearch

```c
/***************************************************************
** Function name:mazeSearch
** Descriptions:前进N格
** input parameters:iNblock(前进的格数)
** output parameters:无
** Returned value:无
***************************************************************/
void mazeSearch(void)
{
    char cL=0, cR=0, cCoor=1;
    if (__GmLeft.cState) {
        cCoor=0;
    }
    /*
     *  设定运行任务
     */
    __GucMouseState=__GOAHEAD;
    __GiMaxSpeed=SEARCHSPEED;
    __GmRight.cDir=__MOTORGOAHEAD;
    __GmLeft.cDir=__MOTORGOAHEAD;
    __GmRight.uiPulse=MAZETYPE*ONEBLOCK;
    __GmLeft.uiPulse=MAZETYPE*ONEBLOCK;
    __GmRight.cState=__MOTORRUN;
    __GmLeft.cState=__MOTORRUN;

    while (__GmLeft.cState!=__MOTORSTOP) {
        if (__GmLeft.uiPulseCtr>=ONEBLOCK) {
                                        /*判断是否走完一格*/
            __GmLeft.uiPulse-=ONEBLOCK;
            __GmLeft.uiPulseCtr-=ONEBLOCK;
            if (cCoor) {
                __mouseCoorUpdate();    /*更新坐标*/
            } else {
                cCoor=1;
            }
        }
        if (__GmRight.uiPulseCtr>=ONEBLOCK) {
                                        /*判断是否走完一格*/
            __GmRight.uiPulse-=ONEBLOCK;
            __GmRight.uiPulseCtr-=ONEBLOCK;
        }
        if (__GucDistance[__FRONT]) {
                                /*前方有挡板,则跳出程序*/
```

```c
            __GmRight.uiPulse=__GmRight.uiPulseCtr + 70;
            __GmLeft.uiPulse=__GmLeft.uiPulseCtr+70;
            while (1) {
                if((__GmLeft.uiPulseCtr + 20)>__GmLeft.uiPulse){
                    goto End;
                }
            }
        }
        if (cL) {                    /*是否允许检测左边*/
            if ((__GucDistance[__LEFT]&0x01)==0) {
                                     /*左边有支路，跳出程序*/
                __GmRight.uiPulse=__GmRight.uiPulseCtr+74;
                __GmLeft.uiPulse=__GmLeft.uiPulseCtr+74;
                while ((__GucDistance[__LEFT]&0x01)==0) {
                    if((__GmLeft.uiPulseCtr+20)>__GmLeft.uiPulse) {
                        goto End;
                    }
                }
                __GmRight.uiPulse=MAZETYPE*ONEBLOCK;
                __GmLeft.uiPulse=MAZETYPE*ONEBLOCK;
            }
        } else {            /*左边有挡板时开始允许检测左边*/
            if (__GucDistance[__LEFT]&0x01) {
                cL=1;
            }
        }
        if (cR) {                    /*是否允许检测右边*/
            if ((__GucDistance[__RIGHT]&0x01)==0) {
                                     /*右边有支路，跳出程序*/
                __GmRight.uiPulse=__GmRight.uiPulseCtr+74;
                __GmLeft.uiPulse=__GmLeft.uiPulseCtr+74;
                while ((__GucDistance[__RIGHT]&0x01)==0) {
                    if ((__GmLeft.uiPulseCtr+20)>__GmLeft.uiPulse) {
                        goto End;
                    }
                }
                __GmRight.uiPulse=MAZETYPE*ONEBLOCK;
                __GmLeft.uiPulse=MAZETYPE*ONEBLOCK;
            }
        } else {
            if (__GucDistance[__RIGHT]&0x01) {
                             /*右边有挡板时开始允许检测右边*/
                cR=1;
            }
        }
    }
}
```

```
End:__mouseCoorUpdate();        /*更新坐标*/
}
```

2）等高图制作程序

智能鼠结合自身法则,整合所有坐标的挡板信息,规划最优路径。

核心函数2：mapStepEdit

```
/***************************************************************
** Function name:mapStepEdit
** Descriptions:制作以目标点为起点的等高图
** input parameters:uiX（目的地横坐标）
**                  uiY（目的地纵坐标）
** output parameters:GucMapStep[][]:各坐标上的等高值
** Returned value:无
***************************************************************/
void mapStepEdit (char   cX, char   cY)
{
    uchar n=0;                    /*有多个支路的坐标的数量*/
    uchar ucStep=1;               /*等高值*/
    uchar ucStat=0;               /*统计可前进的方向数*/
    uchar i,j;

    GmcStack[n].cX=cX;            /*起点X值入栈*/
    GmcStack[n].cY=cY;            /*起点Y值入栈*/
    n++;
    /*
     *  初始化各坐标等高值
     */
    for (i=0; i<MAZETYPE; i++) {
        for (j=0; j<MAZETYPE; j++) {
            GucMapStep[i][j]=0xff;
        }
    }
    /*
     *  制作等高图,直到堆栈中所有数据处理完毕
     */
    while (n) {
        GucMapStep[cX][cY]=ucStep++;           /*填入等高值*/

        /*
         *  对当前坐标格里可前进的方向统计
         */
        ucStat=0;
        if ((GucMapBlock[cX][cY]&0x01)&&       /*前方有路口*/
            (GucMapStep[cX][cY+1]>(ucStep))) {
```

```c
                            /*前方等高值大于计划设定值*/
    ucStat++;               /*可前进方向数加1*/
}
if ((GucMapBlock[cX][cY]&0x02)&&    /*右方有路口*/
    (GucMapStep[cX+1][cY]>(ucStep))) {
                            /*右方等高值大于计划设定值*/
    ucStat++;               /*可前进方向数加1*/
}
if ((GucMapBlock[cX][cY]&0x04)&&
    (GucMapStep[cX][cY-1]>(ucStep))) {
    ucStat++;               /*可前进方向数加1*/
}
if ((GucMapBlock[cX][cY]&0x08)&&
    (GucMapStep[cX-1][cY]>(ucStep))) {
    ucStat++;               /*可前进方向数加1*/
}
/*
 *  没有可前进的方向，则跳转到最近保存的分支点
 *  否则任选一可前进方向前进
 */
if (ucStat==0) {
    n--;
    cX=GmcStack[n].cX;
    cY=GmcStack[n].cY;
    ucStep=GucMapStep[cX][cY];
} else {
    if (ucStat>1) {     /*有多个可前进方向，保存坐标*/
        GmcStack[n].cX=cX;  /*横坐标X值入栈*/
        GmcStack[n].cY=cY;  /*纵坐标Y值入栈*/
        n++;
    }
    /*
     *  任意选择一条可前进的方向前进
     */
    if ((GucMapBlock[cX][cY]&0x01)&&    /*上方有路口*/
        (GucMapStep[cX][cY+1]>(ucStep))) {
                            /*上方等高值大于计划设定值*/
        cY++;               /*修改坐标*/
        continue;
    }
    if ((GucMapBlock[cX][cY]&0x02)&&    /*右方有路口*/
        (GucMapStep[cX+1][cY]>(ucStep))) {
                            /*右方等高值大于计划设定值*/
        cX++;               /*修改坐标*/
        continue;
    }
    if ((GucMapBlock[cX][cY]&0x04)&&    /*下方有路口*/
```

```
                    (GucMapStep[cX][cY-1]>(ucStep))) {
                                /*下方等高值大于计划设定值*/
                cY--;          /*修改坐标*/
                continue;
            }
            if ((GucMapBlock[cX][cY]&0x08)&&    /*左方有路口*/
                    (GucMapStep[cX-1][cY]>(ucStep))) {
                                /*左方等高值大于计划设定值*/
                cX--;          /*修改坐标*/
                continue;
            }
        }
    }
}
```

3）指定坐标跳转程序

该程序块的作用是智能鼠按照最短路径运行到指定坐标点，当然该功能的实现前提是智能鼠已经搜索过该坐标点。

核心函数3：objectGoTo

```
/*****************************************************************
** Function name:objectGoTo
** Descriptions:使智能鼠运动到指定坐标
** input parameters:cXdst（目的地横坐标）
**                  cYdst（目的地纵坐标）
** output parameters:无
** Returned value:无
*****************************************************************/
void objectGoTo (char  cXdst, char  cYdst)
{
    uchar ucStep=1;
    char  cNBlock=0, cDirTemp;
    char  cX,cY;
    cX=GmcMouse.cX;
    cY=GmcMouse.cY;
    mapStepEdit(cXdst,cYdst);    /*制作等高图*/
    /*
     *  根据等高值向目标点运动，直到达到目的地
     */
    while ((cX!=cXdst)||(cY!=cYdst)) {
        ucStep=GucMapStep[cX][cY];
        /*
         *  任选一个等高值比当前自身等高值小的方向前进
         */
        if ((GucMapBlock[cX][cY]&0x01)&& /*上方有路口*/
```

```c
                (GucMapStep[cX][cY+1]<ucStep)) {
                                /*上方等高值较小*/
            cDirTemp=UP;        /*记录方向*/
            if (cDirTemp==GucMouseDir) {
                                /*优先选择不需要转弯的方向*/
                cNBlock++;      /*前进一个方格*/
                cY++;
                continue;       /*跳过本次循环*/
            }
        }
        if ((GucMapBlock[cX][cY]&0x02)&& /*右方有路口*/
            (GucMapStep[cX+1][cY]<ucStep)) {
                                /*右方等高值较小*/
            cDirTemp=RIGHT;     /*记录方向*/
            if (cDirTemp==GucMouseDir) {
                                /*优先选择不需要转弯的方向*/
                cNBlock++;      /*前进一个方格*/
                cX++;
                continue;       /*跳过本次循环*/
            }
        }
        if ((GucMapBlock[cX][cY]&0x04)&& /*下方有路口*/
            (GucMapStep[cX][cY-1]<ucStep)) {
                                /*下方等高值较小*/
            cDirTemp=DOWN;      /*记录方向*/
            if (cDirTemp==GucMouseDir) {
                                /*优先选择不需要转弯的方向*/
                cNBlock++;      /*前进一个方格*/
                cY--;
                continue;       /*跳过本次循环*/
            }
        }
        if ((GucMapBlock[cX][cY]&0x08)&& /*左方有路口*/
            (GucMapStep[cX-1][cY]<ucStep)) {
                                /*左方等高值较小*/
            cDirTemp=LEFT;      /*记录方向*/
            if (cDirTemp==GucMouseDir) {
                                /*优先选择不需要转弯的方向*/
                cNBlock++;      /*前进一个方格*/
                cX--;
                continue;       /*跳过本次循环*/
            }
        }
        cDirTemp=(cDirTemp+4-GucMouseDir)%4;
                                /*计算方向偏移量*/
        if (cNBlock) {
            mouseGoahead(cNBlock);   /*前进cNBlock步*/
```

```
            }
            cNBlock=0;              /*任务清零*/
            /*
             *  控制智能鼠转弯
             */
            switch (cDirTemp) {
            case 1:
                mouseTurnright();
                break;
            case 2:
                mouseTurnback();
                break;
            case 3:
                mouseTurnleft();
                break;

            default:
                break;
            }
    }
    /*
     * 判断任务是否完成，否则继续前进
     */
    if (cNBlock) {
        mouseGoahead(cNBlock);
    }
}
```

4）未搜索支路数统计程序

该程序用于统计指定坐标四周存在的还未搜寻过的支路总数，以供系统运用搜索策略。程序如下：

```
                     核心函数4：crosswayCheck
/****************************************************************
** Function name:crosswayCheck
** Descriptions:统计某坐标存在还未走过的支路数
** input parameters:ucX（需要检测点的横坐标）
**                  ucY（需要检测点的纵坐标）
** output parameters:无
** Returned value:ucCt（未走过的支路数）
****************************************************************/
uchar crosswayCheck (char  cX, char  cY)
{
    uchar ucCt=0;
```

```c
    if ((GucMapBlock[cX][cY]&0x01)&&
                              /*绝对方向,迷宫上方有路口*/
        (GucMapBlock[cX][cY+1])==0x00) {
                              /*绝对方向,迷宫上方未走过*/
        ucCt++;               /*可前进方向数加1*/
    }
    if ((GucMapBlock[cX][cY]&0x02)&&
                              /*绝对方向,迷宫右方有路口*/
        (GucMapBlock[cX+1][cY])==0x00) {
                              /*绝对方向,迷宫右方没有走过*/
        ucCt++;               /*可前进方向数加1*/
    }
    if ((GucMapBlock[cX][cY]&0x04)&&
                              /*绝对方向,迷宫下方有路口*/
        (GucMapBlock[cX][cY-1])==0x00) {
                              /*绝对方向,迷宫下方未走过*/
        ucCt++;               /*可前进方向数加1*/
    }
    if ((GucMapBlock[cX][cY]&0x08)&&
                              /*绝对方向,迷宫左方有路口*/
        (GucMapBlock[cX-1][cY]) == 0x00) {
                              /*绝对方向,迷宫左方未走过*/
        ucCt++;               /*可前进方向数加1*/
    }
    return ucCt;
}
```

5)TQD-Micomouse-JD走迷宫主程序

```c
/*****************************************************************
** Function name:main
** Descriptions:主函数
** input parameters:无
** output parameters:无
** Returned value:无
*****************************************************************/
main (void)
{
    uchar n=0;                /*有多个支路的坐标的数量*/
    uchar ucRoadStat=0;       /*统计某一坐标可前进的支路数*/
    uchar ucTemp=0;           /*用于START状态中坐标转换*/
    mouseInit();              /*底层驱动的初始化*/
    Init_7289();              /*显示模块的初始化*/
    while (1) {
        switch (GucMouseTask) {   /*状态机处理*/
        case WAIT:
```

```c
            sensorDebug();
            voltageDetect();
            delay(100000);
            if (keyCheck()==true) {    /*检测按键等待启动*/
                Reset_7289();          /*复位ZLG7289*/
                GucMouseTask=START;
            }
            break;
        case START:              /*判断智能鼠起点的横坐标*/
            mazeSearch();        /*向前搜索*/
            if (GucMapBlock[GmcMouse.cX][GmcMouse.cY]&0x08) {
                                 /*判断智能鼠左边是否存在出口*/
                if (MAZETYPE==8) {
                                 /*修改四分之一迷宫的终点坐标*/
                    GucXGoal0=1;
                    GucXGoal1=0;
                }
                GucXStart=MAZETYPE-1;
                                 /*修改智能鼠起点的横坐标*/
                GmcMouse.cX=MAZETYPE-1;
                                 /*修改智能鼠当前位置的横坐标*/
            /*
             *由于默认的起点为(0,0)，现在需要把已记录的挡板资料转换过来
             */
                ucTemp=GmcMouse.cY;
                do {
            GucMapBlock[MAZETYPE-1][ucTemp]=GucMapBlock[0][ucTemp];
                    GucMapBlock[0][ucTemp]=0;
                }while (ucTemp--);
            /*
             * 在OFFSHOOT[0]中保存起点坐标
             */
                GmcCrossway[n].cX=MAZETYPE-1;
                GmcCrossway[n].cY=0;
                n++;
                GucMouseTask=MAZESEARCH;
                                 /*状态转换为搜寻状态*/
            }
            if (GucMapBlock[GmcMouse.cX][GmcMouse.cY]&0x02) {
                                 /*判断智能鼠右边是否存在出口*/
            /*
             * 在OFFSHOOT[0]中保存起点坐标
             */
                GmcCrossway[n].cX=0;
                GmcCrossway[n].cY=0;
                n++;
                GucMouseTask=MAZESEARCH;
```

```c
                                    /*状态转换为搜寻状态*/
            }
            break;
        case MAZESEARCH:
            if(((GmcMouse.cX==GucXGoal0)&&(GmcMouse.cY==GucYGoal0))
||((GmcMouse.cX==GucXGoal0)&&(GmcMouse.cY==GucYGoal1))
                ||((GmcMouse.cX==GucXGoal1)&&(GmcMouse.cY==GucYGoal0))
||((GmcMouse.cX==GucXGoal1)&&(GmcMouse.cY==GucYGoal1)))
            {
                mouseTurnback();
                objectGoTo(GucXStart,GucYStart);
                mouseTurnback();
                GucMouseTask=SPURT;
                break;
            }
            else{
                ucRoadStat=crosswayCheck(GmcMouse.cX,GmcMouse.cY);
                                /*统计可前进的支路数*/
                if (ucRoadStat)
                {                  /*有至少一条可前进方向*/
                    if (ucRoadStat>1) {
                                /*有多条可前进方向，保存坐标*/
                        GmcCrossway[n].cX=GmcMouse.cX;
                        GmcCrossway[n].cY=GmcMouse.cY;
                        n++;
                    }
                    crosswayChoice(); /*用右手法则搜索选择前进方向*/
                    mazeSearch();/*前进一格*/
                }
                 else if(ucRoadStat==1)
                 {
                    crosswayChoice();
                            /*用右手法则搜索选择前进方向*/
                    mazeSearch();
                 }
                 else
                 {          /*没有可前进方向，回到最近支路*/
                    mouseTurnback();
                    n=n-1;
                    objectGoTo(GmcCrossway[n].cX,GmcCrossway[n].cY);

                    ucRoadStat=crosswayCheck(GmcMouse.cX,GmcMouse.cY);
                    if (ucRoadStat>1) {
                        GmcCrossway[n].cX=GmcMouse.cX;
                        GmcCrossway[n].cY=GmcMouse.cY;
                        n++;
                    }
```

```c
                crosswayChoice();
                mazeSearch();
            }
        }
        break;
    case SPURT:
        mouseSpurt();       /*以最优路径冲向终点*/
        objectGoTo(GucXStart,GucYStart);   /*回到起点*/
        mouseTurnback(); /*向后转,恢复出发姿势*/
        while (1) {
            if (keyCheck()==true) {
                break;
            }
            sensorDebug();
            delay(20000);
        }
        break;
    default:
        break;
    }
}
```

思考与总结

(1)智能鼠程序设计各个阶段有何联系?

(2)智能鼠主要函数分别起什么作用?

(3)智能鼠走迷宫是由多个主要函数配合执行的,只有都调试准确,才能使智能鼠成功通过迷宫到达终点。

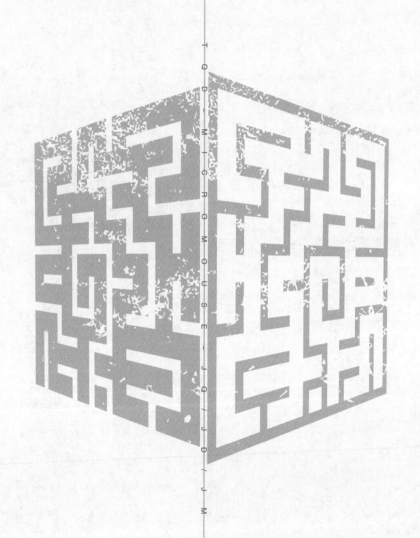

附录A

风靡全球的国际智能鼠走迷宫竞赛

2019年是智能鼠走迷宫竞赛有史以来,最具兴盛发展、硕果累累的一年,在世界各地如火如荼地举行国际智能鼠走迷宫竞赛如图A-1所示。

1月,在印度孟买举办印度智能鼠国际竞赛。

3月,在美国加利福尼亚州举办APEC国际智能鼠竞赛。

4月,在葡萄牙Gondomar(贡多马尔)举办国际智能鼠走迷宫竞赛。

5月,在中国天津举办IEEE智能鼠走迷宫国际邀请赛。

6月,在英国伦敦举办智能鼠国际竞赛。

8月,在智利举办智能鼠走迷宫国际竞赛。

10月,在埃及举办埃及智能鼠国际竞赛。

11月,在日本东京举办全日本智能鼠国际公开赛。

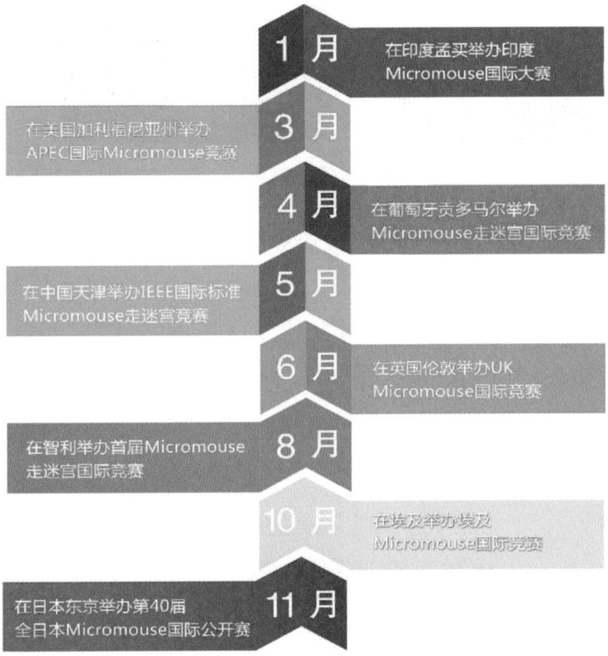

图A-1 国际智能鼠竞赛赛事安排

国际智能鼠走迷宫竞赛将成为全球高等教育、职业教育、普通教育、技术创新产教融合发展的助推器。在人工智能智能鼠走迷宫竞赛蓬勃发展的国际大环境下，教育领域适时地引进国际知名赛事提升学生的专业综合能力，掌握实践与创新的经验，助力产教融合发展，为行业、产业、企业培养更多优秀种子人才。

1. 中国IEEE智能鼠走迷宫国际邀请赛

从2009年开始，天津启诚伟业科技有限公司把智能鼠走迷宫竞赛引入中国，将IEEE智能鼠走迷宫竞赛进行本土化创新改革，对满足产业优化升级，开阔国际视野，掌握实践与创新经验，培育高技术高技能人才，起到了引领推动作用。

从2016年至2019年，连续四届举办中国IEEE智能鼠走迷宫国际邀请赛，该竞赛由天津市教育委员会主办，天津启诚伟业科技有限公司和天津渤海职业技术学院承办，如图A-2所示。

图A-2 从2016年开始举办"中国IEEE智能鼠走迷宫国际邀请赛"

中国IEEE智能鼠走迷宫国际邀请赛

目前，中国IEEE智能鼠走迷宫国际邀请赛设置了"中学、高职、本科、硕士、职业"共五个竞赛组别，旨在提升大赛的社会参与度和专业覆盖面。智能鼠走迷宫竞赛已经发展成为了系统化培养和教育的重要载体。充分体现光机电结合、软硬件结合、控制与机械结合，演绎"工程"课程概念的同时，延伸和扩展了"创新"课程的理念，使得学生的学习内容和教师的授课方式都有了全新的内涵，真正着眼于综合素质的培养，创造快乐素质教育。

中国IEEE智能鼠走迷宫国际邀请赛主要特点：

（1）参赛群体：既面对在校大学生，也面对小学、中学和职业人士，体现贯通式培养，终身教育的特点。还包括国际智能鼠专业级选手和历届国际智能鼠竞赛获奖选手。

（2）迷宫场地：既有面向中小学的8×8智能鼠迷宫场地，也有面向大专

院校的16×16全迷宫古典智能鼠场地；更有面对精英选手的25×32半尺寸智能鼠迷宫场地。体现竞赛的延展性，以智能鼠走迷宫竞赛为核心形式，不同学习阶段的学生都可以参赛。

（3）竞赛项目：既有智能鼠走迷宫赛项，又有自走车赛项，体现了竞赛既有技术性也有工程性，以工程应用为导向的竞赛思想。

（4）竞赛规则：普通教育、职业教育、高等教育、职业精英竞赛规则的相同点和差异点对比见表A-1。

表A-1 竞赛规则相同点和差异点对比

参赛类别	普通教育组	职业教育组	高等教育组	职业精英组
竞赛形式	（1）程序参数APP在线调试。 （2）图形化趣味编程。 （3）IOT智能传感技术应用。 （4）8×8迷宫竞速	（1）理论知识考核。 （2）根据裁判现场任务编程并实现相应功能。 （3）现场技术答辩。 （4）16×16古典迷宫竞速	（1）DIY外观及结构机械设计。 （2）硬件技术创新。 （3）程序算法创新。 （4）16×16古典迷宫竞速	（1）DIY外观及结构机械设计。 （2）硬件技术创新。 （3）程序算法创新。 （4）25×32半尺寸迷宫竞速
竞赛内容	（1）组装任务10%。 （2）调试任务40%。 （3）竞速任务50%	（1）理论考核20%。 （2）创新赛30%。 （3）竞速赛50%	（1）创新赛20%。 （2）竞速赛80%	竞速赛100%

智能鼠国际专家现场培训指导如图A-3所示。

图A-3 智能鼠国际专家现场培训指导

2. 美国APEC世界智能鼠竞赛

1977年在美国纽约举行的首场令人震撼的智能鼠走迷宫竞赛，由IEEE与APEC共同主办。于是诞生了国际上最有影响力的美国APEC世界智能鼠竞赛。号称智能鼠世界三大赛事之一，截止到2019年已经举办了34届。

APEC组织的官网网址：http://www.apec-conf.org/。

美国智能鼠爱好者的网址：http://micromouseusa.com/，如图A-4所示。

图A-4 美国APCE世界智能鼠竞赛官网截屏

竞赛时间：每年2月到4月之间。

竞赛地点：每年不同（举办过的地点包括北卡罗来纳州、德克萨斯州、佛罗里达州、加利福尼亚州等），每年都会有来自美国、英国、日本、韩国、新加坡、印度、中国等国家的选手踊跃参赛，如图A-5所示。

图A-5 中国选手参加第30届美国APEC世界智能鼠竞赛

3. 英国智能鼠国际竞赛

从1980年至今，英国智能鼠国际竞赛已经成长为国际知名的智能鼠竞赛之一。

竞赛时间：每年6月。

竞赛地点：英国伯明翰城市大学。

该项竞赛由英国智能鼠和机器人协会主办，英国的智能鼠竞赛特点在于重在参与，从中学生、大学生、到社会人员，任何人都可以参赛。所有的参赛队员分为不同的组别，迷宫难度也适当调整。该项竞赛分为line follower、wall follower、maze solver等项目，吸引了来自全球10余个国家，50余支队伍参赛。

英国智能鼠国际竞赛评分规则介绍：在16×16的迷宫中，参赛智能鼠需要完成起点到终点的搜索和全迷宫的遍历，求解最佳路线并完成由起点到终点的

冲刺。计分时间=搜索时间（第一次搜索到终点的时间）/30+冲刺时间（完成起点到终点最短路径的高速冲刺）+惩罚时间（撞挡板罚时：3 s/次）。

官方网址：https://ukmars.org/index.php/Main_Page，如图A-6所示。

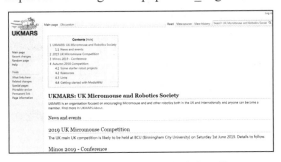

图A-6　英国智能鼠国际竞赛官网截屏

4. 全日本智能鼠国际公开赛

全日本智能鼠国际公开赛从1980年到2019年已经举办40届。

竞赛时间：每年的11月底或12月初。

竞赛地点：日本东京。

官方网址：http://www.ntf.or.jp/mouse/Micromouse2018/index.html，如图A-7所示。

图A-7　全日本智能鼠国际公开赛官网截屏

每年竞赛都有来自美国、英国、日本、新加坡、中国、蒙古、智利、葡萄牙等二十多个国家的智能鼠参赛队角逐（见图A-8）。

赛项由古典智能鼠赛项、半尺寸智能鼠赛项、自走车赛项组成。参赛队由中学生、大学生和职业精英组成，据统计有300多支参赛队。全日本智能鼠国际公开赛可以说是代表当今，国际智能鼠技术领域级别最高、技术最强的赛事，所以备受瞩目。

图A-8　第39届全日本智能鼠国际公开赛颁奖照片

5. 智利智能鼠走迷宫国际竞赛

智利外交部希望通过国际智能鼠走迷宫竞赛，推动智利青少年科技创新以及国际间技术创新和交流合作，从而带动智利经济发展。2018年12月3日在全日本智能鼠国际公开赛期间，智利驻日本大使馆主持召开"智利智能鼠走迷宫国际竞赛研讨会"特别邀请国际智能鼠专家（美国的David Otten、英国的Peter Harrison、日本的中川友纪子、中国的宋立红、智利的Benjamin等）共同商议智利智能鼠走迷宫国际竞赛统一标准和规范，如图A-9、图A-10所示。

图A-9　智利外交部会议——共同探讨智能鼠发展

图A-10　2018年智利智能鼠走迷宫国际竞赛规则研讨会

6. 葡萄牙智能鼠走迷宫国际竞赛

2019年4月27日在Gondomar（贡多马尔）举办，由葡萄牙杜罗大学技术执行委员会主办。

葡萄牙竞赛开始于2011年，旨在通过培养创造力和能力来提供完整的技术学习环境，迄今已经成功举办9届。

竞赛时间：每年4月或5月。

竞赛地点：葡萄牙。

官方网址：http://www.micromouse.utad.pt/，如图A-11所示。

2019年4月27日，当地时间18时整，在葡萄牙波尔图体育馆中，来自英国、中国、葡萄牙、西班牙、巴西、新加坡等国家的参赛队，正在上演一场紧张激烈的国际智能鼠走迷宫竞赛。伴随着中国智能鼠稳健搜索和极速冲刺，掌声欢呼声在葡萄牙波尔图体育馆雷鸣般响起……启诚智能鼠实现突破性成果，取得了世界亚军的殊荣（见图A-12）。

图A-11　葡萄牙智能鼠走迷宫国际竞赛官网截屏

图A-12　启诚智能鼠荣获葡萄牙智能鼠走迷宫竞赛世界亚军

赛后葡萄牙智能鼠走迷宫竞赛组委会主席安东尼奥表示，近年来中国的综合国力和技术实力不断增强，特别是教育领域对于科技创新和工程素养越来越重视。启诚智能鼠首次参加葡萄牙智能鼠走迷宫竞赛，就获得了优异成绩非常可喜可贺（见图A-13）。

图A-13　中国及葡萄牙智能鼠走迷宫竞赛专家现场技术交流

视　频

葡萄牙Micromouse走迷宫国际竞赛

7. 印度智能鼠国际竞赛

2020年1月4日，印度孟买举办"2020年第23届亚洲科技节首届智能鼠国际大赛"，来自印度、中国、澳大利亚、尼泊尔、斯里兰卡、孟加拉等国家的代表队参加本届大赛（见图A-14）。中国天津智能鼠代表队力克群雄，以绝对优势包揽"金、银、铜"全部奖牌，将17.5万卢比奖金完美收入囊中。

竞赛时间：每年1月份。

竞赛地点：印度孟买。

官方网址：http://techfest.org/competitions/Micromouse。

图A-14 印度智能鼠国际竞赛合影

特别值得一提的是，印度金奈理工学院鲁班工坊代表队（见图A-15）采用2017年中方赠送的IEEE国际标准智能鼠走迷宫创新型教学设备TQD-Micromouse-JD智能鼠参加本届大赛，荣获印度国内大赛冠军、世界精英组第四名的好成绩，并赢得5 000卢比奖金，成为印度国际智能鼠走迷宫竞赛的明星赛队。印度鲁班工坊指导教师卡西克表示，金奈理工学院鲁班工坊代表队能取得这样优异的成绩，是三年来鲁班工坊的师生和支持企业（启诚科技）共同努力的成果。

图A-15 印度鲁班工坊参赛师生合影

8. 埃及智能鼠国际竞赛

埃及国际电气电子工程师学会（IEEE）现在已经发展成为具有较大影响力的国际学术和技术组织之一。30多年来，一直在推动和指导电气电子技术的发展与创新。这项技术包括电子元件、电路理论和设计技术的应用，以及针对有效转换、控制和电力状况分析工具的开发。IEEE成员包括杰出的研究人员、从业人员和杰出的获奖者。

图A-16所示为埃及IEEE在官网首页为智能鼠竞赛做的宣传。IEEE电力电子和可再生能源大会为颇具亮点的国际智能鼠大赛优胜者准备了丰厚的奖金。特等奖相当于1 000美元；杰出表现奖相当于700美元；最佳创新设计奖相当于500美元。参赛队来自埃及国内或国际工程学或相关专业的学生，也可以是高中生。每个参赛队中最多允许有两名学生。

官方网址：http://www.ieee-cpere.org/International_Competition.html。

图A-16　埃及智能鼠国际竞赛

埃及智能鼠国际竞赛纪实如图A-17所示。

图A-17　埃及智能鼠国际竞赛纪实照片

附录B

进阶级经典竞赛案例分析

IEEE国际标准智能鼠走迷宫竞赛具有一定难度,是一项富有挑战性和趣味性的学生比赛,在国内外享有一定的知名度和影响力。智能鼠走迷宫竞赛项目,从技术上涵盖了物联网应用技术、电子信息工程技术、嵌入式技术、通信技术、软件技术、计算机网络技术、信息安全技术、移动通信技术、计算机应用技术、应用电子技术、计算机控制技术、机电一体化技术、自动化技术等多个专业技术,涉及传感器检测、人工智能、自动控制和机电运动部件应用等技能和综合职业素养。全面展现高等教育和职业教育的发展水平,可提高电子信息类高素质、高技能应用型人才的培养质量。

通过竞赛,推动了电子信息行业企业人才能力需求,顺应科技发展将嵌入式技术开发、智能算法优化等前端先进技术融入竞赛内容中,进一步深化校企合作,引导电子信息类专业开展单片机应用、嵌入式技术应用、物联网技术应用等。课程建设和教学改革,促进创新型人才培养模式的发展,增强了电子信息类专业学生就业竞争力,推进创新创业教育,强化创业指导和服务,提高就业水平。

随着时代的发展,科技的进步,智能鼠顺应现代科技发展,经过多年的蜕变与优化,已经成为集人工智能、嵌入式、智能传感等融合新技术于一体的优秀实训教育平台。

四十多年来,IEEE每年举办一次国际性的智能鼠走迷宫竞赛。自举办以来,各个国家和地区的学生踊跃参加,尤其是美国和欧洲国家的高校学生,为此有些大学还特别开设了"智能鼠原理与制作"的选修课程,如图B-1、图B-2所示。

下面就以比较有代表性的国内外智能鼠走迷宫竞赛典型迷宫地图为例进行解析说明。

1. 美国APEC世界Micromouse竞赛迷宫分析

2015年3月16日,第30届美国APEC世界Micromouse竞赛成功举办。天津启诚伟业科技有限公司带领天津大学生联合代表队参加本次竞赛,让天津智能鼠技术与国际接轨,是天津智能鼠历史性的转折,具有里程碑式的意义。

图B-1　组委会主席David Otten教授

图B-2　日本选手宇都宫正和

本次届竞赛是一届非常成功的赛事。传感器从数字型向模拟型过渡，运动结构也从步进电动机向空心杯直流电动机发展，在这次赛事上出现了融合吸地风扇的智能鼠。

本次竞赛所采用的迷宫整体难度上相对均衡，有众多的路径可供选择，既有长直道展现智能鼠的高速运动性能，也有适中的连续转弯体现智能鼠的精确控制，如图B-3所示。

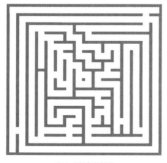
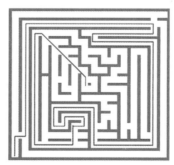

（a）迷宫原图　　　　　　　　　（b）最佳路径

图B-3　迷宫图解析

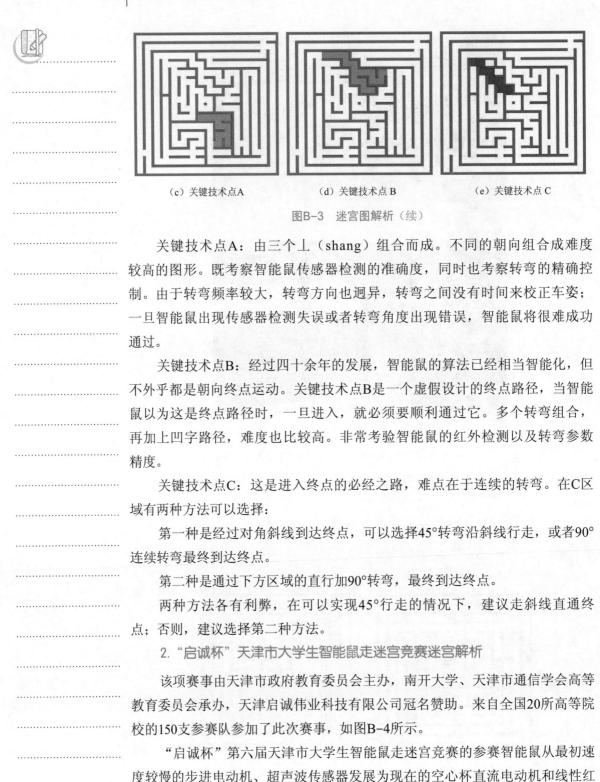

(c) 关键技术点 A　　　　　(d) 关键技术点 B　　　　　(e) 关键技术点 C

图 B-3　迷宫图解析（续）

关键技术点 A：由三个上（shang）组合而成。不同的朝向组合成难度较高的图形。既考察智能鼠传感器检测的准确度，同时也考察转弯的精确控制。由于转弯频率较大，转弯方向也迥异，转弯之间没有时间来校正车姿；一旦智能鼠出现传感器检测失误或者转弯角度出现错误，智能鼠将很难成功通过。

关键技术点 B：经过四十余年的发展，智能鼠的算法已经相当智能化，但不外乎都是朝向终点运动。关键技术点 B 是一个虚假设计的终点路径，当智能鼠以为这是终点路径时，一旦进入，就必须要顺利通过它。多个转弯组合，再加上凹字路径，难度也比较高。非常考验智能鼠的红外检测以及转弯参数精度。

关键技术点 C：这是进入终点的必经之路，难点在于连续的转弯。在 C 区域有两种方法可以选择：

第一种是经过对角斜线到达终点，可以选择 45° 转弯沿斜线行走，或者 90° 连续转弯最终到达终点。

第二种是通过下方区域的直行加 90° 转弯，最终到达终点。

两种方法各有利弊，在可以实现 45° 行走的情况下，建议走斜线直通终点；否则，建议选择第二种方法。

2. "启诚杯" 天津市大学生智能鼠走迷宫竞赛迷宫解析

该项赛事由天津市政府教育委员会主办，南开大学、天津市通信学会高等教育委员会承办，天津启诚伟业科技有限公司冠名赞助。来自全国 20 所高等院校的 150 支参赛队参加了此次赛事，如图 B-4 所示。

"启诚杯" 第六届天津市大学生智能鼠走迷宫竞赛的参赛智能鼠从最初速度较慢的步进电动机、超声波传感器发展为现在的空心杯直流电动机和线性红

外传感器。参赛选手的水平也在逐年提高。在2017年举办的竞赛中,采用的是一个非常经典的迷宫,难度高、开放性强是它最主要的特点,如图B-5所示。

图B-4 2017年"启诚杯"智能鼠竞赛开幕式

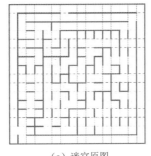 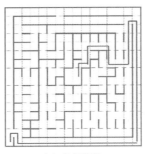

(a) 迷宫原图　　　　　　　(b) 最佳路径

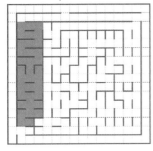 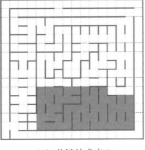 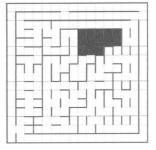

(c) 关键技术点A　　　(d) 关键技术点B　　　(e) 关键技术点C

图B-5 迷宫图解析

关键技术点A:由于第一个路口紧贴起点位置,使用智能算法的智能鼠通常都不会在这里右转,这也就造成了智能鼠沿着最外侧挡板行走一整圈后又回到了这一路口,最终进入A区域。几乎完全对称的图案,众多的路口,使得智能鼠几乎没有机会进行车姿校正。这对智能鼠的性能是一个巨大的考验。

关键技术点B:与关键技术点A相同,B区域也是以高开放性为主。众多

的路口，使得智能鼠没有足够的时间进行车姿校准，同时又给了高性能智能鼠45°走斜线的机会。智能鼠的性能差距，很容易判断出参赛选手水平高低。

关键技术点C：C区域是进入终点的必经之路，使用智能算法的智能鼠进入这一区域后，通常不会去行走"死胡同"，直接进入终点完成竞赛。这里也成为区分算法是否智能的区域。

3. 第二届IEEE智能鼠走迷宫国际邀请赛迷宫解析

2017年8月，第二届"中国IEEE智能鼠走迷宫国际邀请赛"成功举办。本届大赛吸引了来自英国、泰国、蒙古等国家代表队，以及中国实力雄厚的天津大学、南开大学、北京交通大学、天津中德应用技术大学等智能鼠精英赛队参加，如图B-6所示。

图B-6　第二届IEEE智能鼠走迷宫国际邀请赛

本次竞赛采用了一幅比较有特色的迷宫。随着智能算法的流行，大型竞赛所选用的迷宫，越来越注重为智能算法增加难度系数，如图B-7所示。

关键技术点A：蓝色箭头所在位置是智能算法的首选路口。这一部分是一个完全封闭的区域，难度也较低；但是当智能鼠进入这一区域，就需要面临大量的无用搜索和转弯。只有成功搜索完毕并顺利退出A区域，智能鼠才能进入其他区域的搜索。

关键技术点B：B区域包含智能鼠到达终点的两条路径。大量的路口以及左转右转结合，非常注重考察智能鼠红外检测精度和转弯角度的准确性。智能鼠在任何一个路口发生传感器误判或者转弯精度有误都将是致命的。

关键技术点C：C区域同样是到达终点的一条路径。阶梯状的连续转弯，非常考验智能鼠的转弯准确度。在可以实现45°斜直线行走的情况下，推荐选择这一路径。

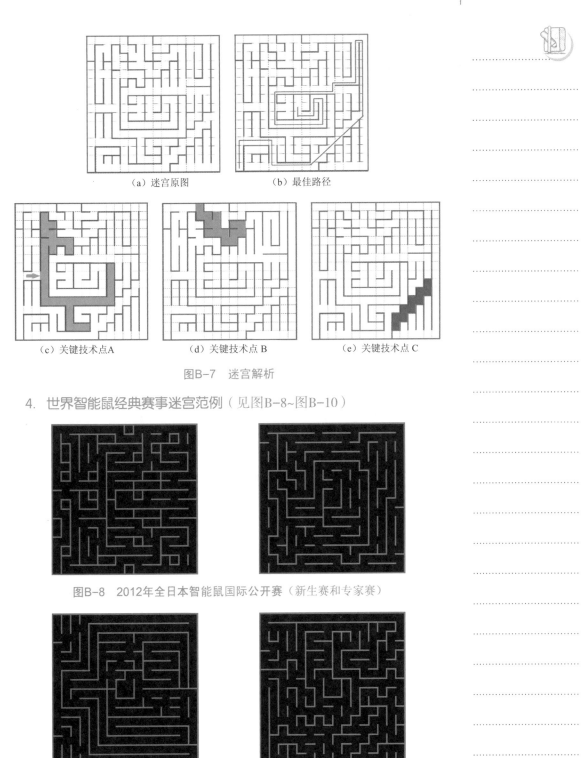

(a) 迷宫原图　　(b) 最佳路径

(c) 关键技术点 A　　(d) 关键技术点 B　　(e) 关键技术点 C

图 B-7　迷宫解析

4. **世界智能鼠经典赛事迷宫范例**（见图 B-8~图 B-10）

图 B-8　2012 年全日本智能鼠国际公开赛（新生赛和专家赛）

图 B-9　2000 年英国 UK 智能鼠
　　　　国际邀请赛

图 B-10　2002 年美国 APEC
　　　　智能鼠国际大赛

附录C TQD-Micromouse-JD器件清单

TQD-Micromouse-JD器件清单见表C-1。

表C-1 器件清单

序号	名称	数量	备注
1	TQD-Micromouse-JD	1	
2	专用充电器	1	
3	专用下载器	1	
4	专用下载器连接线	1	
5	USB线	1	
6	专用电池	1	
7	配套光盘	1	

附录D 教学内容和学时安排

本课程参考教学学时为48学时,具体分配表见表D-1。

表D-1 学时分配

序号	内容	学时
第一篇 基础知识篇	项目一 智能鼠的发展历程 项目二 智能鼠的硬件结构 项目三 智能鼠的开发环境 项目四 智能鼠的基本功能操作	20
第二篇 综合实践篇	项目一 智能鼠高级功能 项目二 智能鼠实战任务	14
第三篇 拓展竞技篇	项目一 智能鼠的路径规划和行为决策算法 项目二 智能鼠路径规化的原理 项目三 智能鼠走迷宫程序设计	14
总计		48

附录E 电路图形符号对照表

电路图形符号对照表如表E-1所示。

表E-1 电路图形符号对照表

序号	名称	国家标准的画法	软件中的画法
1	发光二极管		
2	二极管		
3	电解电容元件		
4	接地		
5	按钮开关		
6	可调电阻元件		

附录F 专业词汇中英文对照表

专业词汇中英文对照表见表F-1~表F-3。

表F-1 与智能鼠相关的专业词汇

中文	英文	中文	英文
核心控制模块	main control module	左方、左斜、前方、右斜、右方	the left, the front-left, the front, the front-right and the right
主控芯片	main control chip	g段	the g segment
输入模块	input module	步进电动机	stepping motor
输出模块	output module	电动机驱动电路	motor drive circuit
核心板电路	main control circuit	真值表	truth table
电源电路	power circuit	H桥电路	H-bridge circuit
控制电路	control circuit	转动(步进电动机)	rotate
外围电路	peripheral circuit	电子元器件	electronic component
键盘显示电路	keyboard-display circuit	晶振	crystal oscillator
JTAG接口电路	JTAG interface circuit	电容	capacitance
按键电路	key-pressingcircuit	限流可调电阻	adjustable current-limiting resistance
数据传输	data transmission	数码管	digitron
人机交互系统	human-computer interaction system	外围器件	peripheral device
红外传感器	IR sensor	脉冲振荡电路	pulse oscillation circuit
红外检测电路	infrared detection circuit	脉冲信号	pulse signal
红外线	infrared light	方波	square wave
红外校准	infrared calibration	感知系统	perceptual system
红外强度	infrared intensity	载波频率	carrier frequency
红外发射头	infrared transmitter	原理图	schematic diagram
红外接收头	infrared receiver	软件界面	software interface
PWM信号发生器模块	PWM signal generator driver module	驱动库	driver library

表F-2 与竞赛相关的专业词汇

中文	英文	中文	英文
单元格	cell	路口	crossing
挡板	wall	电子自动计分系统	electronic automatic scoring system
立柱	post	参赛队员	competitor
竞赛场地	competition maze	智能鼠竞速比赛	Micromouse competition
起点	the start	最优路径	the optimal path
目的地/终点	the destination	轨迹	trajectory
迷宫坐标	the coordinate in the maze	通道	passage way

表F-3 与智能算法相关的专业词汇

中文	英文	中文	英文
底层驱动	the bottom driver program	路径规划和决策算法	path planning and decision algorithm
顶层算法	the top algorithm program	结构体	struct
算法	algorithm	差速控制	differential-speed control
策略	strategy	直线运动	straight movement
法则（左、右手法则）	rule	转弯	turning
右手、左手、中心法则	the right-hand rule, the left-hand rule, the central rule	校正车姿	correct the attitude
90°、180°转弯	90-degree turning/180-degree turning	运行校正	attitude correction
编程并实现	programming and realizing	核心函数	core function
等高图	step map	（驱动步进电动机的）时序状态	time sequence status
循环检测	cycle detection	前进一格	moving forward one cell
"8字型"路径运行控制	movement control in picture-8-shaped path	按键等待	waiting for button press
实现避障	obstacle avoidance	判断车姿	determining the attitude
运动姿态的控制	motion attitude control	暂停一步	waiting one step
两轮差速	two-wheel difference speed	精确转弯控制	accurate turning control

附录G "智能鼠原理与制作"国际实训课程标准

(适用于高等职业学校实训课程)

一、适用专业

电子信息类专业、计算机类专业、通信类专业、自动化类专业等。

包含：电子信息工程技术（专业代码610101）、应用电子技术（专业代码610102）、智能产品开发（专业代码610104）、智能终端技术与应用（专业代码610105）、电子测量技术与仪器（专业代码610112）、物联网应用技术（专业代码610119）、计算机应用技术（专业代码610201）、软件技术（专业代码610205）、嵌入式技术与应用（专业代码610208）、物联网工程技术（专业代码610307）、机电一体化技术（专业代码560301）、电气自动化技术（专业代码560302）、工业过程自动化技术（专业代码560303）、智能控制技术（专业代码560304）、工业网络技术（专业代码560305）、工业自动化仪表（专业代码560306）、工业机器人技术（专业代码560309）。

二、课程性质

"智能鼠原理与制作"从技术上涵盖了物联网应用技术、电子信息工程技术、嵌入式技术、软件技术、计算机应用技术、应用电子技术、机电一体化技术、自动化技术、智能控制技术等多种专业，涉及传感器检测、人工智能、自动控制和机电运动部件应用等技能和综合职业素养。在实训中主要学习智能鼠的硬件结构、智能鼠的开发环境、智能鼠的红外检测、智能鼠的运动姿态控制、智能鼠的路径规划和行为决策法则，重点考查学生在电工电子技术、单片机技术、电机控制技术、嵌入式技术及程序设计基础知识，与工程实践创新类相关的基本知识及基本能力。

本课程的教学内容是采用项目式教学模式，遵循递进原则，从"玩转"到"掌握"，再到"精通"，丰富学习者的工程实践知识和技术应用经验，拓展学习者的专业视野，内化形成良好的职业素养，提升学习者的实践创新能力。

本课程开设在二年级第一学期，前序课程为单片机控制技术、C语言程序设计、电工电子技术和传感器与检测技术，后续课程为自动控制技术和面向对象程序设计。

三、课程目标

本课程以智能鼠的硬件结构、智能鼠的开发环境、智能鼠的红外检测、智能鼠的运动姿态控制、智能鼠的路径规划和行为决策法则等项目为专业核心能力。采用"教、学、做"一体化的方式，完成专业能力、社会能力、方法能力的培养。

1. 专业能力

（1）掌握常用仪器仪表的使用方法。
（2）掌握智能鼠装配、调试的基本原理与方法。
（3）掌握嵌入式开发环境的基本使用方法。
（4）掌握对整体项目进行程序设计的方法。
（5）掌握传感信号分析与处理的方法。
（6）掌握步进电动机控制的方法。
（7）掌握在迷宫中实现智能搜索与路径规划的相关知识。
（8）掌握智能鼠设备调试过程中的常见故障分析及排除方法。

2. 方法能力

（1）能独立进行资料与信息的收集与整理。
（2）能独立制订与实施工作计划。
（3）具备探究能力和运用理论知识解决实际问题的能力。

3. 社会能力

（1）具有沟通及团队协作的能力。
（2）具有勇于创新、敬业、乐业的工作作风。
（3）具有安全意识、质量意识与责任意识。

四、设计思路

共设计了5个项目。在项目中，明确了各项目的任务目标、任务内容、教师知识与能力要求、学生知识与能力准备、教学材料、实施步骤以及完成该项目所需的学时的内容，项目教学全过程按引入→搜集→目标→实施→测试→评估六步法来组织进行。

建议总学时：48学时。总学分：3学分。

五、内容大纲

（1）适用对象：三年制高职学生。

（2）参考学时：48学时。

（3）学习目标：完成智能鼠硬件系统结构分析与组装、开发环境的搭建、传感与检测信号调试、直行及转弯控制与姿态矫正、智能搜索与路径规划的实现。

（4）工作项目：

项目一　智能鼠的硬件结构（6学时）

项目描述：初步认识精典智能鼠结构组成；分析机械结构设计原理，认识精典智能鼠各个部分电子元器件以及其工作原理、特性和安装方式；完成智能鼠硬件组装。

项目目标：掌握核心控制器功能和特点；熟悉基础常用仪器仪表测试方法，尝试测量相关电路模块（电路通断、VCC和GND等）；按照说明书或相关图纸学习精典智能鼠的装配。

项目二　智能鼠的开发环境（4学时）

项目描述：熟悉智能鼠开发环境——国际开源IAR软件的安装和使用；通过IAR软件调试程序，并对智能鼠进行程序下载。

项目目标：熟悉软件安装方法、操作方法；熟悉相关驱动的安装方法；学会正确连接精典智能鼠、下载器与计算机并下载程序。

项目三　传感与检测信号调试（10学时）

项目描述：了解精典智能鼠装置上传感器的工作原理及使用方法。了解扩展模块数码管和按键的作用。

项目目标：学习按键和数码管的使用；掌握智能鼠装置传感器在电路板上的分布原理；通过调试过程，学习使用传感器测距并通过数码管显示，实现多传感器协同工作。

项目四　智能鼠的运动姿态控制（14学时）

项目描述：了解精典智能鼠的运动结构——步进电动机的控制原理，完成智能鼠直行与转弯控制。

项目目标：掌握智能鼠步进电动机的驱动方式，并学习速度调节和转弯角度调节；学习如何通过传感器修正智能鼠的动作。

项目五　智能鼠的路径规划和行为决策法则（14学时）

项目描述：了解智能鼠的智能搜索与路径规划的相关算法；实现智能鼠走迷宫路径规划。

项目目标：学习迷宫坐标定义及其挡板信息的存储方法；学会智能鼠迷宫路径的规划方法——等高图；熟练使用左右手法则、中心法则搜索并记忆迷宫；学习智能鼠走迷宫竞赛的程序结构，最终找到最优路径。

六、技能考核要求

理论和实际操作考核。

总评成绩以百分制计算，分为平时成绩考核和期末综合考核两部分。平时成绩考核一般为出勤、项目作业、测验，占总成绩的30%；期末综合考核一般为理论考试、实际操作考核，占总成绩的70%，其中理论考试占总成绩的30%，实际操作考核占40%。

七、实施建议

（1）教师应依据工作任务中的典型产品为载体安排和组织教学活动。

（2）教师应按照项目的学习目标编制项目任务书。项目任务书应明确教师讲授（或演示）的内容；明确学习者预习的要求；提出该项目整体安排以及各模块训练的时间、内容等。如以小组形式进行学习，对分组安排及小组讨论（或操作）的要求，也应做出明确规定。

（3）教师应以学习者为主体设计教学结构，营造民主、和谐的教学氛围，激发学习者参与教学活动，提高学习者学习积极性，增强学习者学习信心与成就感。

（4）教师应指导学习者完整地完成项目，并将有关知识、技能与职业道德和情感态度有机融合。

八、教学条件

智能鼠创新创客实训室，具有可以指导教学的专业教材。

九、学习评价

教师评价：对学生每一个实践环节都进行评价和反馈。

学生评价：学生分组完成任务后，抽出两个课时来让他们进行互相学习和评价，便于在以后的环节中改善。

十、教材和推荐资料

1. 推荐教材

王超, 高艺, 宋立红. 智能鼠原理与制作：进阶篇[M]. 北京：中国铁道出版社有限公司，2019.

2. 参考资料

［1］黄智伟. 32位ARM微控制器系统设计与实践[M]. 北京：北京航空航天大学出版社, 2010.

［2］来清民, 来俊鹏. ARM Cortex-M3嵌入式系统设计和典型实例[M]. 北京：北京航空航天大学出版社, 2013.

十一、说明

编写人：姚　嵩　刘宝生　宋立红　范平平　李　萌　陈立考　邱建国

审核人：龚　威　高　艺　康晓明

2020年3月24日

Intelligent Micro Motion Device (Micromouse) Technology and Application Series
Research Result of Tianjin "the Belt and Road" Joint Laboratory (Research Center)
Engineering Practice Innovation Project (EPIP) Teaching Mode Planned Textbook

Micromouse Design Principles and Production Process
(Intermediate)

Compiled by Wang Chao, Gao Yi, Song Lihong
Translated by Wang Juan, Fan Pingping, Yan Jingyi

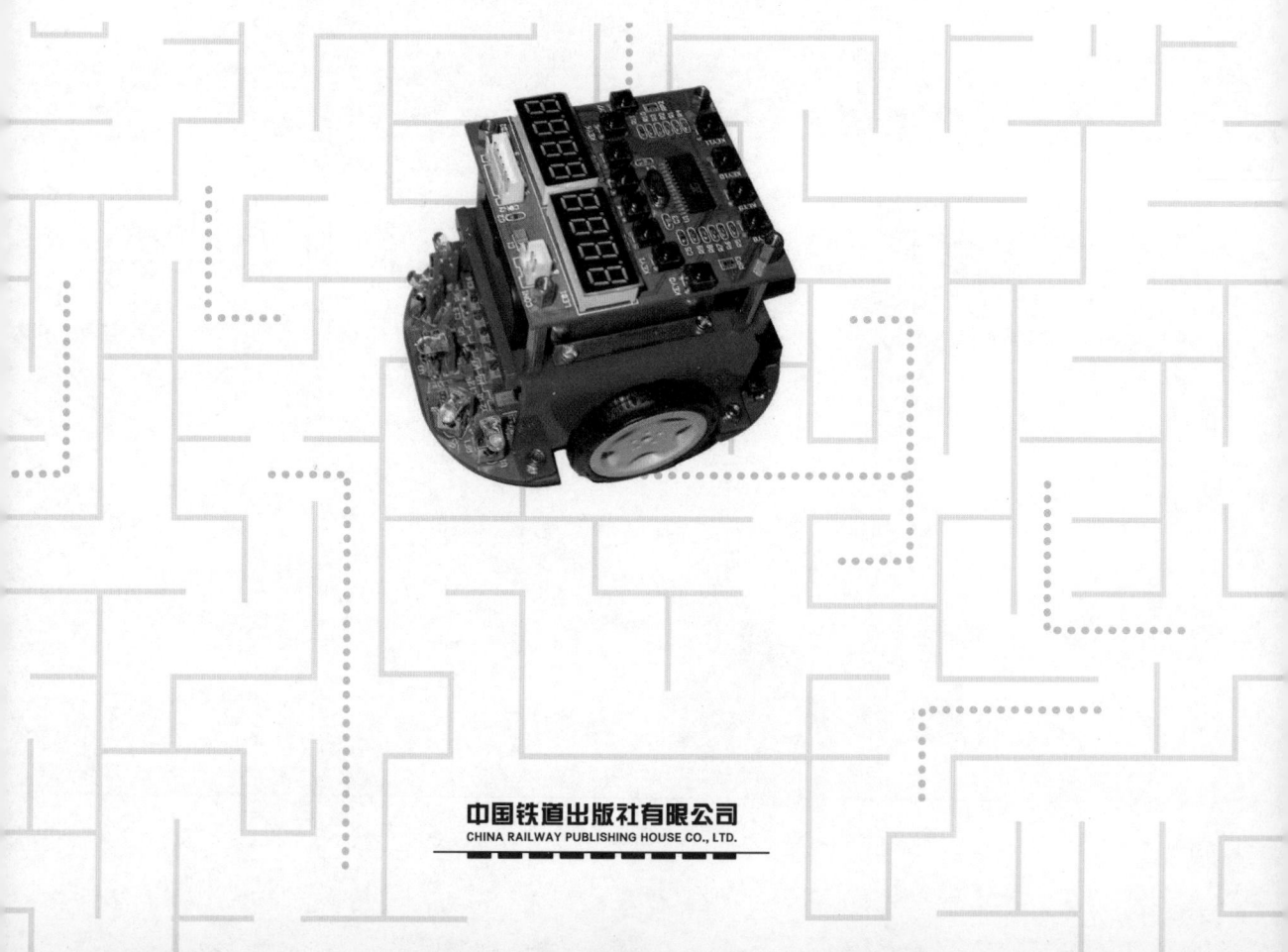

中国铁道出版社有限公司
CHINA RAILWAY PUBLISHING HOUSE CO., LTD.

Introduction to the contents

The book is bilingual in both Chinese and English, based on the TQD-Micromouse-JQ provided by Tianjin Qicheng Science and Technology Co., Ltd., which is the intermediate-level of a series of books on Micromouse Technology and Application. The book is based on real engineering projects, through "Elementary Knowledge", "Comprehensive Practice" and "Advanced Skills and Competitions". All of these three chapters are describe the development, hardware, development environment, and function debugging of the Micromouse; advanced function of Micromouse, actual combat task of Micromouse and path planning of Micromouse, etc. The appendix of this book provides the relevant knowledge of the international Micromouse maze competition, such as the device list of TQD-Micromouse-JQ, the Micromouse maze library, bilingual comparison table of glossary, and the international curriculum standard.

The book is equipped with a wealth of resources such as videos, pictures, texts, etc. on important knowledge points, skills points and literacy points. The reader can obtain relevant information by scanning the QR code in the book.

The book is suitable as the textbook of vocational colleges and universities corresponding professional comprehensive and innovative practice. It can also be used as a training book for relevant engineering and technical personnel or reference book for Micromouse lovers.

About the Authors

Wang Chao

Wang Chao is currently a Professor in School of Electrical and Information Engineering, Tianjin University, China. He is a member of the Teaching Guiding Committee for Automation Majors under the Ministry of Education of China. His current research interests include multiphase flow measurement and instrumentation, electrical tomography (ERT, ECT, EMT and EST). His courses at Tianjin University include computer control technology and industrial control networks. Since 2010, he has introduced Micromouse as an important carrier of practical teaching into school of electrical and information engineering for the first time. Two teams of Tianjin University won the first and second place at 2018 APEC Micromouse competition.

Gao Yi

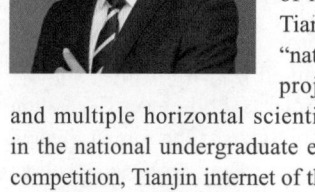

Gao Yi is a master supervisor of School of Electronic Information Engineering at Nankai University, deputy director of Electronic Information Experimental Teaching Center, deputy director of Youth Working Committee of Tianjin MCU Society and member of the judging group of many technology competition of college students and Tianjin vocational skills competition. He has participated in "national high-tech research and development plan (863 plan) projects", "Tianjin science supporting plan key projects" and multiple horizontal scientific research projects. He has led teams to participate in the national undergraduate electronic design competition, Tianjin electronic design competition, Tianjin internet of things competition, Tianjin IEEE Micromouse competition for college students, APEC international Micromouse competition and national robot competition.

Song Lihong

Song Lihong is general manager of Tianjin Qicheng Science and Technology Co., Ltd. and founder of Qicheng Micromouse. The company has been committed to the research and development, design, production, promotion and service work of teaching instruments about embedded system, internet of things and artificial intelligence used in higher education, vocational education and basic education. Qicheng has sponsored the "Qicheng Cup" Micomouse maze competition of college students and the intelligent micro motion device (Micromouse) competition of vocational college skills competition more than 40 times. Since 2016, the company has been actively engaged in the technical support service of international project Luban Workshop. As an innovative educational equipment in China, Qicheng Micromouse has been brought to Thailand, India, Indonesia, Pakistan, Cambodia, Nigeria, Egypt and other countries and has been favored by teachers and students in these countries. Qicheng Micromouse has made contributions to "the Belt and Road" initiative.

Wang Juan

Deputy director and associate professor of International Cooperation and Exchange Department of Tianjin Light Industry Vocational Technical College. Since 2016, as a chief participant, completed the construction of Luban Workshops in India and Egypt. Guiding student to take part in the Tianjin Oral English Contest and won the first prize. Also guiding student to win the third prize in the First Higher Vocational Oral English Contest of Tianjin Haihe Education Park, and win the Excellent Prize in the Global Brand Planning Contest  in China. Participating in the teaching achievement award project, won the first prize of national teaching achievement, the special prize and the second prize of Tianjin Teaching Achievement Award. Since 2017, presided over and participated in a number of research projects. And participated in the translation of the international textbook *Detecting for Assembling and Adjusting of NC Machine Tools.*

Fan Pingping

Lecture of Electronic Information and Automation Institute of Tianjin Light Industrial Vocational Technical College, Director of Electric Automation. The owner of the second prize of the National Vocational College Information Teaching Design competition and the second prize of the National Vocational College Teachers' Micro Class competition. Excellent Instructor of Tianjin Vocational College Student Skills Competition and National  Vocational College Skills Competition. Guiding students to get awards in Computer Mouse Maze, Installation and Debugging of Wind and Solar Hybrid Power Generation System, University Robot Competition, Tianjin University Students Innovation Method Application Competition, Intelligent Elevator Installation and Maintenance Competition, etc. Main lecture of National New Energy Professional Teaching Resource Library Course of "SCM Control Technology". Main lecture of Tianjin University's new era "Course Ideological and political reform" of "The Application of SCM Technique".

Yan Jingyi

Yan Jingyi, assistant to the general manager of Tianjin Qicheng Science and Technology Co.,Ltd., studied in the University of California, Santa Cruz. In 2015, she served as the accompaniment interpreting for professor MIT David Otten, chairman of the organizing committee of APEC Micromouse international competition. In 2018, she served as the English translator and simultaneous interpreter for Mr. Davaanyam, vice chairman of the board of directors of  new Mongolia Education Group when he visited in Tianjin. In 2018, she went to Cambodia as a volunteer interpreter for Bun Phearin, president of National Polytechnic Institute of Cambodia. She was helping Cambodia students learn the Chinese Micromouse technology at the same time. Yan makes contributions to the education in countries along "the Belt and Road" for a long time.

FOREWORD

Micromouse is a micro, intelligent motion device (or embedded microrobot) composed of embedded micro controllers, sensors and electromechanical moving parts. Micromouse can reach the predetermined destination fast by automatically memorizing all the routes in various mazes and selecting the optimal path out with the aid of suitable algorithms. The Micromouse competition involves wide-ranging scientific knowledge such as mechatronics, cybernetics, optics, programming and artificial intelligence.

For more than 40 years, the Institute of Electrical and Electronics Engineers (IEEE) has hosted an annual Micromouse competition. Since its inception, the international event has featured extensive and active participation, especially that of students from colleges and universities in the US and Europe. Some universities even offer an elective course on the principles and production process of Micromouse. In 2007, Shanghai and some other cities in the Yangtze River Delta started to stage small-scale, experimental Micromouse competitions. In 2009, Tianjin Qicheng Science and Technology Co., Ltd. introduced the competition to Tianjin and added local features to create an updated version based upon the Engineering Practice Innovation Project (EPIP) teaching model. These pioneering efforts helped boost future Micromouse events and played a key role in integrating them into classroom teaching. Years of exploration and progress have made Micromouse competitions to serve as educational platforms that encourage innovation and practice. By bringing together expertise and interest, the multi-dimensional and pioneering competitions have been essential for cultivating students' capabilities in practice and innovation, reforming curriculum and improving education.

In order to further promote and apply the achievements of Micromouse, we specially organized the book of *Micromouse Design Principles and Production Process* (*Intermediate*) for vocational-technical learning and training. This book is based on the TQD-Micromouse-JD provided by Tianjin Qicheng Science and

Technology Co.,Ltd. As a teaching carrier, practice teaching is carried out from shallow to deep, and from easy to difficult, step-by-step teaching.

This book will teach users with basic principles of Micromouse and end up as a profession in the field by following a step-by-step method that is also used in compiling the book. After reading the book, users will acquire more knowledge about engineering practice, enrich their experience in technology application, open up a broader vision in expertise, and become more professional.This book's ultimate goal is to cultivate innovation-minded practitioners. The select cases in the book are all adapted from real-life engineering projects. Also all the authors are came from enterprises, colleges and universities that have long remained committed to R&D on Micromouse or that have won awards in international competitions.

This book offers abundant videos, pictures and texts, etc. on important knowledge points, skills points and literacy points. The reader can scan the QR code in the book to get these supporting resources. The authors' rich international teaching experience has made the book became a practical teaching carrier to promote international talent training. Vocational colleges and institutions of higher education can use the book to guide their practicums on innovation. The book is suitable as the textbook of vocational colleges and universities corresponding professional comprehensive and innovative practice. It can also be used as a training book for relevant engineering and technical personnel or reference book for Micromouse lovers.

The appendix of this book, the international curriculum standard for "Micromouse Design Principles and Production Process" (applicable to higher vocational colleges) is provided. This course defines the following aspects as the core of students' professional competencies: the hardware structure, the development environment, the infrared detection, the movement and attitude control, the path planning and behavior decision-making algorithm of Micromouse. The method combining "teaching, learning and experimenting" is adopted in order to build students' professional competences, social ability and methodological ability. The content of this course is highly integrated

FOREWORD | III

with the "Luban workshop" construction projects in many countries. This course provides China educational standards and provides rich practical teaching resources for all countries along "the Belt and Road" route, serving the training of skilled personnel in various fields. Intelligent micro motion device (Micromouse) technology and application series are the research result of Tianjin "the Belt and Road" Joint Laboratory (Research Center)— Tianjin Sino-German and Cambodia Intelligent Motion Device and Communication Technology Promotion Center and are also the Engineering Practice Innovation Project (EPIP) teaching model planned textbook. Amid the ongoing efforts to build emerging engineering programs, relevant universities can also use the book to guide the integration of IT and automation technology and innovative teaching models.

  This book is co-authored by Wang Chao, professor of Tianjin University; Gao Yi, associate professor of Nankai University; and Song Lihong, general manager of Tianjin Qicheng Science and Technology Co.,Ltd., the founder of Qicheng Micromouse. The English version was translated and compiled by associate professor Wang Juan, lecturer Fan Pingping of Tianjin Light Industry Vocational Technical College, and Yan Jingyi, a general manager assistant of Tianjin Qicheng Science and Technology Co., Ltd. Associate professor Liu Baosheng, lecturer Li Meng of Tianjin Transportation Technical College and Wang Danyang, a lecturer of Tianjin Light Industry Vocational Technical College are participated in some translation work of this book. David Otten, professor of Massachusetts Institute of Technology (MIT) in USA, Peter Harrison, professor of Birmingham City University in UK and António Valente, professor of University of Trás-os-Montes and Alto Douro in Portugal are proofreaders of the English version and they specially wrote congratulatory letters for this books. The book has received great and generous support from scholars and experts who come from Tianjin University, Nankai University, Tianjin Light Industry Vocational Technical College, Tianjin Transportation Technical College, MIT, Birmingham City University, and University of Trás-os-Montes. Chen Likao, Qiu Jianguo and Song Shan, are the employees of Tianjin Qicheng Science

and Technology Co.,Ltd. provided practical engineering cases, QR codes, videos and PPT course resources for the book. We also owe great gratitude to Tianjin Municipal Education Commission, China Railway Publishing House Co., Ltd. and Tianjin Qicheng Science and Technology Co.,Ltd. for their invaluable guidance and support. This book is sponsored compiled by Tianjin Light Industry Vocational Technical College, published by China Railway Publishing House Co., Ltd. and will be used in countries along "the Belt and Road" route through the Luban Workshop program.

There will be some gaps or even errors in the book due to a tight publishing schedule and insufficient consideration from the authors, so any constructive criticisms and suggestions are greatly welcomed.

Authors
August, 2020

Contents

Chapter 1 Elementary Knowledge..................................001

Project 1 Evolution of Micromouse... 003

Task 1 Origins of Micromouse 003

Task 2 Competition and Debugging Environment of Micromouse........ 010

Project 2 Micromouse Hardware Structure 014

Task 1 Components of Micromouse.. 014

Task 2 Main Control Board of Micromouse.. 016

Project 3 Development Environment of Micromouse................................ 017

Task 1 IAR EWARM Development Environment 017

Task 2 Downloading Program of Micromouse....................................... 018

Project 4 Basic Function Control of Micromouse 020

Task 1 Human-Computer Interaction System of Micromouse.............. 020

Task 2 Infrared Detection of Micromouse... 023

Task 3 Attitude Control System... 029

Chapter 2 Comprehensive Practice...............................039

Project 1 Advanced Control Function.. 041

Task 1 Straight Movement Control 041

Task 2 Accurate Turning Control 046

Project 2 Actual Combat Tasks 050

Task 1 Using of Debugging Board 7289 050

Task 2 Non-Contact Start/Stop Control... 052

Task 3 Movement Control in Picture-8-Shaped Path 054

Chapter 3 Advanced Skills and Competitions 057

Project 1 Path Planning and Decision Algorithm 059

Task 1 Common Strategies of Maze Searching....................................... 059

Task 2　The Basic Rules of Maze Searching 060

Project 2　The Principle of Path Planning 065

Task 1　The Information Storage Method of a Maze 065

Task 2　The Step Map Making Method 066

Project 3　Micromouse Program Design 068

Task 1　Attitude Program Control 068

Task 2　Analysis of the Main Program Structures 070

Appendix 083

Appendix A　Micromouse Competition Going Popular in the World 085

Appendix B　Analysis of Improved-level Classic Competition Cases 097

Appendix C　Device List of TQD-Micromouse-JD 105

Appendix D　Teaching Content and Class Arrangement 105

Appendix E　The Circuit Diagram Symbol Comparison Table 106

Appendix F　Bilingual Comparison Table of Glossary 107

Appendix G　The International Curriculum Standard for "Micromouse

　　　　　　Design Principles and Production Process" 109

Chapter 1

Elementary Knowledge

The Micromouse competitions have enjoyed worldwide popularity for over four decades. Micromouse is required to search the entire maze without human manipulation to find the destination. And then, Micromouse needs to select, among the many possible paths, the optimal path to reach the destination and spurt from the start to the destination as quickly as possible. Contestants are ranked by the search time plus the spurt time of Micromouse. Mazes used in competitions comply with the international standards set by the Institute of Electrical and Electronics Engineers (IEEE). In this chapter, you will gain a systematic understanding of the Micromouse technology from international standard mazes of IEEE, hardware systems and software development environment of Micromouse. You will also learn in more detail the fundamental principles and practical operations of Micromouse.

Micromouse Design Principles and Production Process (Intermediate)

Chapter 1　Elementary Knowledge　003

Project 1

Evolution of Micromouse

Learning objectives

(1) Learning about the evolution of Micromouse.

(2) Understanding the Micromouse competition platform, i.e. mazes and automatic scoring system.

Task 1　Origins of Micromouse

1. Birth of Micromouse

In 1938, Claude Elwood Shannon, an American mathematician born in Michigan state, completed his master's thesis *A Symbolic Analysis of Relay and Switching Circuits*. He used the Boolean algebra that happens to correspond with the binary system of 0 and 1 to process the relay switches of information in a pulse mode. The notable work had transformed the design of digital circuits both theoretically and technologically, making it an epoch-making thesis in the modern history of digital computers.

In 1948, Shannon published another famous work that is still relevant today, *A Mathematical Theory of Communication*, which earned him the title "Father of Information Theory".

In 1956, Shannon attended the Dartmouth Conference and became one of the founding fathers of the emerging discipline of artificial intelligence. He pioneered the application of artificial intelligence in computer chess and invented a mechanical mouse that could run through a maze autonomously, which proved that computers could improve their intelligence through learning.

2. Evolution of Micromouse in the World

In 1972, *Journal of Mechanical Design* started a contest where mechanical mouse solely driven by mousetrap springs competed with other entries to see which one could cover the longest distance.

In 1977, *IEEE Spectrum* introduced the concept of Micromouse, which is a small robotic vehicle controlled by microprocessors and has the capabilities to decode and navigate in complex mazes.

In 1979, the IEEE initiated a Micromouse competition through its magazine (*Spectrum and Computer*) and it rewarded the designer of the champion Micromouse that could find a way out of the maze all on its own in the shortest possible time span with USD 1,000.

In 1980, the first All Japan Micromouse International Competition was held, followed with more such events, such as UK Micromouse competition in 1980, Singapore IES Micromouse Competition in 1987, and Micromouse Competition for College Students held by China Computer Federation (CCF) in 2007. As shown in Fig.1-1-1.

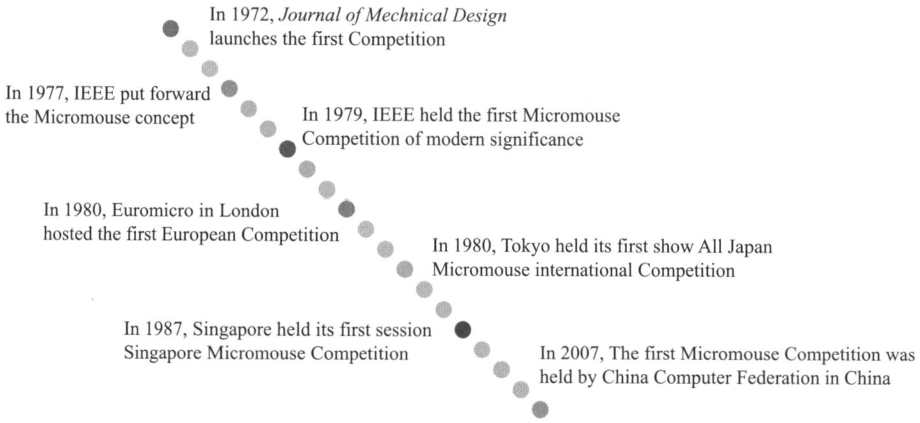

In 1972, *Journal of Mechnical Design* launches the first Competition

In 1977, IEEE put forward the Micromouse concept

In 1979, IEEE held the first Micromouse Competition of modern significance

In 1980, Euromicro in London hosted the first European Competition

In 1980, Tokyo held its first show All Japan Micromouse international Competition

In 1987, Singapore held its first session Singapore Micromouse Competition

In 2007, The first Micromouse Competition was held by China Computer Federation in China

Fig. 1-1-1 Global development trajectory

The past four decades have witnessed the great evolution of Micromouse from the mechanical mouse in 1972 to Micromouse nowadays. The competitions now feature wider participation at all education levels from around the globe. When the competitions were first launched, only graduate students from world-renowned colleges and universities such as Harvard and MIT were able to participate. Later on, students from research universities, universities of applied

Chapter 1　Elementary Knowledge | 005

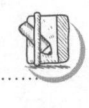

sciences and vocational schools could compete as well. And nowadays even primary and middle school students may take part in Micromouse competitions. Micromouse has been adopted as a teaching vehicle at educational institutions of various levels to cultivate students' engineering literacy, improve the awareness of innovation and boost their design skills.

Micromouse competitions in various forms are flourishing across the world. Now they have grown into global innovation events that are applicable to students at different education levels.

3. Evolution of Micromouse in China

Micromouse competitions have experienced over ten years of growth since 2007 in China, as shown in Fig.1-1-2. In 2007, Tianjin Qicheng Science and Technology Co., Ltd. first introduced the competition to Tianjin and added local features to create an updated version with the advanced Engineering Practice Innovation Program model as a core vision. These pioneering efforts helped boost future Micromouse events in China and played a key role in integrating relevant technologies into classroom teaching.

First Competition	Higher Education	Vocational Education	General Education	International Competition
In 2007, The First Micromouse Competition was held in China	Form 2009, Micromouse Competition has always been a university students' subject competition	Form 2010, Micromouse Competition has always been a colleges students' skills competition	Form 2016, General Education and Vocational Education Micromouse Challenge Contest started to be held in China	Form 2016, International Invitational Micromouse Competition is held every year in China

Fig. 1-1-2　Development trajectory in China

The competitions have helped upgrade industries, broaden the horizon, accumulate experience in practice and innovation and cultivate high-caliber, high-tech and highly skilled personnel (see Fig.1-1-3). A variety of Micromouse competitions has been held in China, such as contests for university students, competitions for vocational colleges, and international challenge competitions for general-vocational high schools, which enable us to gather rich experience and solid technical prowess.

The past decade has witnessed China constantly exploring new ways to make the local Micromouse competitions gain increasing international exposure.

Video

Evolution of Micromouse in China

006 | Micromouse Design Principles and Production Process (Intermediate)

When the competitions were first introduced into China, we simply copied foreign models, but as we gain more experience and build novel platforms for international exchanges and cooperation, foreign countries are also learning from us. Generally speaking, there are three stages in the development of Micromouse in China: imitation and learning; innovation and growth; and going global as a leader.

Fig. 1-1-3   Students at Micromouse competitions

The first stage: In 2015, the Tianjin team went to the United States to participate in the 30th APEC Micromouse Contest and ranked sixth globally (see Fig.1-1-4). In 2017 and 2018, Tianjin Qicheng Technology sponsored the champion team that won the enterprise designated-topic session at Tianjin College Students Micromouse Competition to go to Tokyo and compete in the 38th and 39th All Japan Micromouse International Competition (see Fig.1-1-5). The travel and boarding fees of the contestants were fully covered by the company. The two competitions in Japan enabled us to learn more about global advanced technologies of Micromouse and make connections with industry experts and leaders, greatly boosting the development of Micromouse technologies in China.

Fig. 1-1-4   Tianjin team in the US for APEC 30th Annual Micromouse Contest

• Video

Imitation and learning

Chapter 1　Elementary Knowledge　007

Fig. 1-1-5　Tianjin team at All Japan Micromouse International Competition

　　The second stage: Micromouse competitions were added local, innovative features and underwent necessary reforms to comply with China's realities. A wide range of tiered teaching platforms based upon the TQD-Micromouse produced by Tianjin Qicheng Technology were created to meet the needs of students at different levels: junior high, senior high, undergraduate, and postgraduate. Since 2016, IEEE Micromouse International Invitational Competition in China has featured more extensive participation by world-famous scholars and experts, domestic and foreign teachers and students, and elite teams in China. The name list includes Professor David Otten from MIT, Professor Su Jinghui from Lunghwa University of Science and Technology in Taiwan, China, Professor Huang Mingji from Ngee Ann Polytechnic in Singapore, Professor Peter Harrison from Birmingham City University, and Mr. Yoko Nakagawa, Secretary-General of the Organizing Committee of All Japan Micromouse International Competition; faculty and students from Luban Workshop in Thailand, India, Indonesia, Pakistan and Cambodia; and competition teams from Tianjin, Beijing, Henan and Hebei, to name a few (see Fig.1-1-6, Fig.1-1-7). By signing up for the competitions held in China, international contestants learned more about the Chinese standards, rules, models and philosophy and later accepted them. In this way, global exchanges and collaboration was facilitated and both sides had something meaningful to learn from each other.

Video

Innovation
and growth

　　The third stage: Educational opening-up is integral to China's reform and opening-up initiative. As "One Belt and One Road" initiative gains momentum, the Luban Workshop programme has been launched since 2016 under the guidance of the Ministry of Education. Micromouse, an exemplar of China's excellent teaching aid, has gone global thanks to the programme.

008 | Micromouse Design Principles and Production Process (Intermediate)

● Video

Going global
as a leader

Since then, Tianjin Qicheng Technology has gone to a raft of foreign countries such as Thailand, India, Indonesia, Pakistan, Cambodia, Nigeria and Egypt to promote Micromouse competitions and offer training sessions free of charge, which are well-received by both the local teachers and students(see Fig.1–1–8-Fig.1–1–13). Micromouse has served as a bridge connecting China with the rest of the world!

Fig. 1–1–6   2018 Third IEEE Micromouse International Invitation

Fig. 1–1–7   2019 "Qicheng Cup" IEEE Micromouse International Invitation

Fig. 1–1–8   Micromouse training session at Luban Workshop in India

Chapter 1　Elementary Knowledge　009

Fig. 1–1–9　Micromouse training session at Luban Workshop in Thailand in 2016

Fig. 1–1–10　Micromouse training session at Luban Workshop in Indonesia in 2017

Fig. 1–1–11　Micromouse training session at Luban Workshop in Pakistan in 2018

Fig. 1–1–12　Micromouse training session at Luban Workshop in Cambodia in 2018

Fig. 1–1–13　Micromouse training session at Luban Workshop in Egypt in 2020

Task 2　Competition and Debugging Environment of Micromouse

1. Competition maze

At present, the international Micromouse competition adopts IEEE standard and uses the same specification maze, that is, a square maze composed of 8×8 cells.The "walls" of the maze can be inserted, so that a variety of mazes can be formed.

The TQD-Micromouse Maze 8×8 is shown in Fig.1–1–14. The floor of the maze is 2.96 m×2.96 m, and there are 8×8 standard maze cells on it. Wall and post of the classical Micromouse maze is shown as Fig.1–1–15.

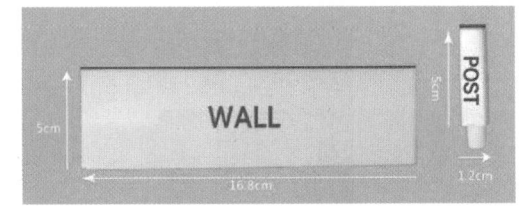

Fig. 1–1–14　TQD-Micromouse　　　Fig. 1–1–15　Wall and post of the classical
　　　　　Maze 8×8　　　　　　　　　　　　　　　Micromouse maze

TQD-Micromouse maze 8×8 specifications are as follows:

(1) The maze is composed of 8×8 square cells with the size of 18 cm×18 cm.

(2) The height of the walls are 5 cm and their thickness are 1.2 cm, so the actual distance of the passageways are 16.8 cm, and the walls seal the whole maze.

(3) The side of the walls are white, and the top are red. The floor of the maze is painted in black. It is made of wood, finished with non-gloss black paint.

Chapter 1　Elementary Knowledge　011

The paint on the side and top of the wall can reflect infrared light, and the floor can absorb infrared light.

(4) The start can be set at one of the four corners. The start must have three walls and only one exit. The destination is located at the center of the maze, which is composed of four cells.

(5) There are small posts, each 1.2 cm×1.2 cm×5 cm, can be inserted at the four corners of each cell. The position of the posts are called lattice points. There are at least one wall to a lattice point except for the destination.

(6) The dimensional accuracy error of the maze making should be no larger than 5%, or less than 2 cm. The joint of the maze floors shall not be more than 0.5 cm, and the gradient change of the joint point shall not be more than 4°. The gap between the walls and posts shall not be more than 1 mm.

(7) The start and the destination shall be designed based on IEEE Micromouse competition rules and standards, that is, Micromouse starts in a clockwise direction.

2. Special testing site

There are 13 marked positions painted on the special testing site and different colors are used to distinguish them(see Fig.1–1–16). They are used for aiding adjusting infrared and turning parameters. Next, let's to learn them:

(1) ① to ②, gray passageways, which is used to detect the offset of Micromouse in the absence of infrared calibration.

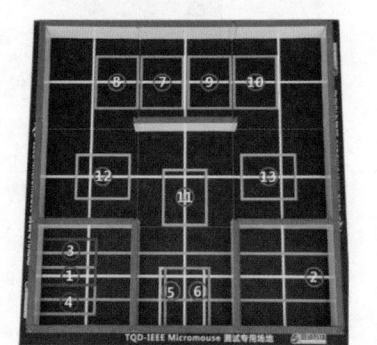

Fig. 1–1–16　Special testing site for TQD-IEEE Micromouse

(2) ③ dark red rectangle, ④ orange rectangle; ③ to ②, ④ to ② are both used to check Micromouse's forward going condition with infrared calibration.

(3) ⑤ yellow rectangle is used to adjust the left front infrared intensity of Micromouse , ⑥ green rectangle is used to adjust the right front infrared intensity of Micromouse ; Correct the attitude.

(4) ⑦, ⑧ green rectangles are used to adjust the right rear infrared intensity of Micromouse , ⑨, ⑩ green rectangles are used to adjust the left rear infrared intensity of Micromouse ; Detect the intersection.

(5) ⑪、⑫、⑬ three blue rectangles are used to debugging 90-degree turning of Micromouse.

3. Automatic scoring system

In order to accurately measure the time used by Micromouse to complete the competition, it is necessary to calculate the time of Micromouse passing the start and the destination full-automatically. The electronic automatic scoring system designed and produced by Tianjin Qicheng Science and Technology Co., Ltd., which is specially used for Micromouse competition, is shown in Fig.1–1–17.

TQD-Micromouse Timer V2.0 system includes the start infrared detection module, the destination infrared detection module, the scoring system module, and the scoring software, etc.

• Video

Working
principle of
TQD-Micro-
mouse imer

Fig. 1–1–17   TQD-Micromouse Timer V2.0

The start infrared detection module and the destination infrared detection module are charged through mini USB. Through a set of inner-placed thru-laser sensors, it can detect the passing of Micromouse. The scoring system module is used to receive the data sent by the infrared detection modules through ZigBee. After been processed by the scoring software in computer, the running condition of Micromouse in the maze is shown in a visualized manner. The scoring software can also be used alone. The start event and destination event can be input through mouse. The overall timing accuracy of the scoring system can reach 0.001 s.

The start infrared detection module and the destination infrared detection module are respectively installed in the start cell and the destination cell, as shown in Fig.1–1–18, Fig.1–1–19. When Micromouse passes by, laser beam is blocked, thus generates a start or destination signal.

Chapter 1 Elementary Knowledge | 013

Fig. 1–1–18　The start

Fig. 1–1–19　The destination

Reflection and Summary

(1) What are the components of the Micromouse maze based on IEEE standard?

(2) What are the characteristics of Micromouse competition?

(3) The usage automatic scoring system has improved the accuracy of competition result calculation greatly. Please briefly explain it's working principle.

014 | Micromouse Design Principles and Production Process (Intermediate)

Project 2

Micromouse Hardware Structure

Learning objectives

(1) Understanding the basic hardware structure of Micromouse.

(2) Learning about the application of Micromouse CPU.

TQD-Micromouse-JD (see Fig. 1–2–1) training-teaching Micromouse is an intelligent robot used for training, teaching and competition. It is independently researched, developed, designed and produced by Tianjin Qicheng Science and Technology Co., Ltd. according to the requirements of practical teaching of Micromouse. It can be used for practical teaching courses of vocational technical colleges, and also as an entry-level platform for the IEEE International Standard Micromouse Competition. It is also a first choice platform for various intelligent robot competitions.

Fig. 1–2–1　TQD-Micromouse-JD

Task 1　Components of Micromouse

TQD-Micromouse-JD hardware composition and advantages:

(1) The MPU uses LM3S615 or STM32, based on the Cortex-M3 ARM core. It has the advantages of fast operation speed, fast interrupt response, and rich peripherals, which ensures that Micromouse has a high intelligence. We also provide a rich library of functions. As long as you can understand C language, you can develop it. Difficulties of using Micromouse are greatly reduced.

(2) TQD-Micromouse-JD is 12 cm long and 9 cm wide. It is short and pithy. It has five sets of infrared digital sensors to detect the walls of the maze. It can

• Video

Hardware
structure of
Micromouse

Chapter 1　Elementary Knowledge　015

also be flexible to complete the 90 degree and 180 degree turns in the maze.

(3) It is driven by stepping motors. It runs smoothly and does not need reduction gear. The mechanical structure is simple. It is especially suitable for beginners' learning application.

(4) On the CPU control board, there are Start key, Reset key, and a 10-pin JTAG debugging interface. There are also 6 GPIOs, a serial port and a SPI interface reserved for the users' free expanding.

(5) Equipped with a charger and rechargeable Li-on battery of 2,200 mA · h, 7.4 V.

(6) LM LINK USB JTAG debugger is easy to use, supports online debugging, and provides maximum convenience for students to program design.

The hardware structure of TQD-Micromouse-JD is shown in Fig. 1–2–2.

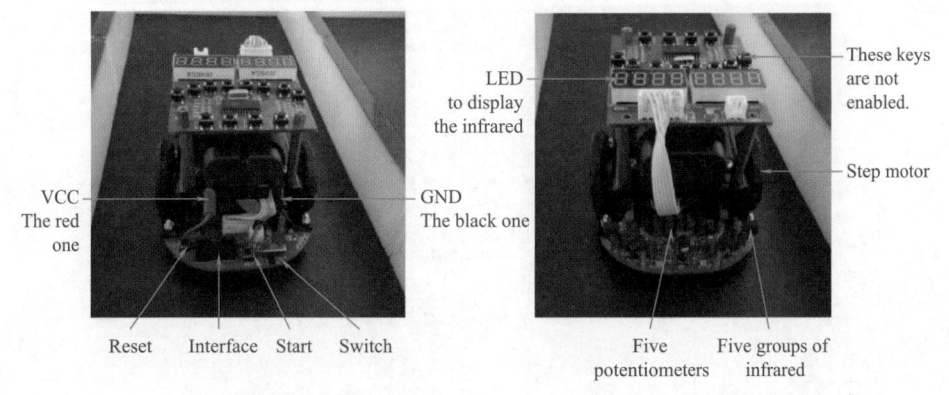

Fig. 1–2–2　TQD-Micromouse-JD hardware structure

The circuit composition diagram of TQD-Micromouse-JD is shown in Fig. 1–2–3, which is mainly composed of main control module, input module and output module. The main control module mainly includes main control circuit, power circuit and control circuit. In addition, the main control circuit extends part of peripheral circuits to form input/output module, which transmits and receive signals during the operation of Micromouse. It includes keyboard-display circuit, JTAG interface circuit, motor-drive circuit, key-pressing circuit and infrared detection circuit.

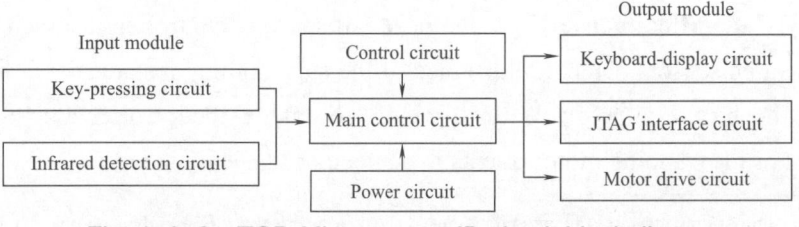

Fig. 1–2–3　TQD-Micromouse-JD circuit block diagram

016 | Micromouse Design Principles and Production Process (Intermediate)

Task 2　Main Control Board of Micromouse

The Micromouse main control board is the core of the whole Micromouse control, which is equivalent to the CPU of Micromouse. it is composed of a LM3S615(QFP-48N) and a few peripheral devices.

The main control board circuit is shown in Fig. 1–2–4. The main body is LM3S615, and the rest are peripheral circuits. Among them, crystal oscillator Y1, capacitance C27 and C28 form pulse oscillation circuit, sending pulse signals to pin 9 and pin10 of the microcontroller.

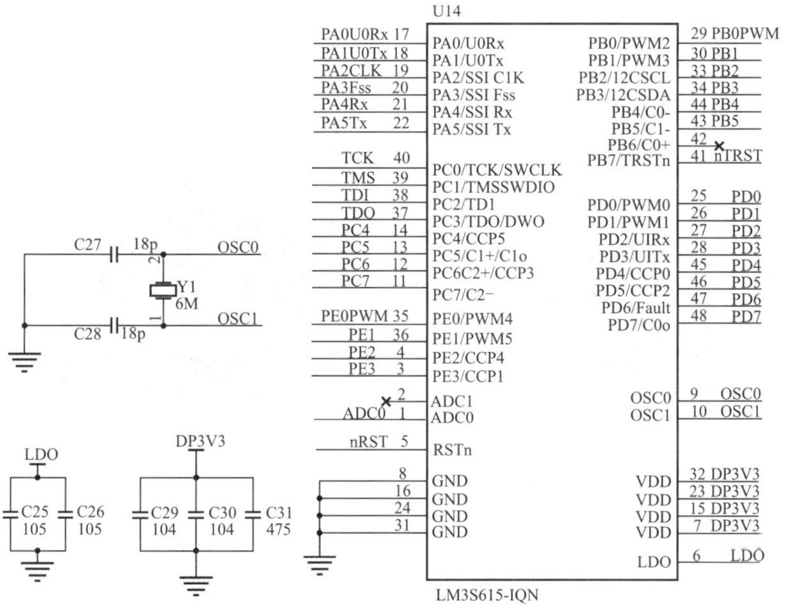

Fig. 1–2–4　The main control circuit [①]

Reflection and Summary

(1) How is data transmission among Micromouse modules?

(2) Are there any similar electronic components that can be used in Micromouse?

(3) TQD-Micromouse-JD is composed of sensors, controllers and actuators. The infrared sensors detect the distance of the surrounding obstacles and display it visually by the digitron. With this infrared data, the controller controls the moving of the stepping motors, so as to achieve obstacle avoidance.

---

① Similar drawings are schematic diagrams derived from Protel 99SE, and their graphic symbols are inconsistent with the national standard symbols. Please refer to the Appendix E for the comparison between the two.

Chapter 1　Elementary Knowledge　017

Project 3

Development Environment of Micromouse

Learning objectives

(1) Learn how to install IAR EWARM.

(2) Learn how to download program to Micromouse.

Task 1　IAR EWARM Development Environment

TQD-Micromouse-JD uses IAR Embedded Workbench for ARM (IAR for short in the following) as the program development environment. It contains project manager, editor, C/C++compiler and ARM assembler, connector XLINK and C-SPY,debugging tool supporting RTOS. Embedded applications program can be easily developed using C/C++ in an EWARM environment. Compared to other ARM development environments, IAR EWARM is easier to get started and used.

We provide a complete driver library and a whole maze demo, including the bottom driver program, top algorithm program and basic experimental procedures; students can develop it only need to know C language.

The software interface is shown in the Fig. 1−3−1.

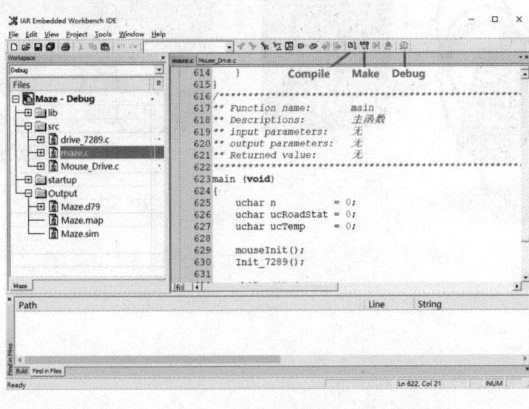

Fig. 1−3−1　IAR software interface

Sofeware

Development environment of Micromouse

Video

Downloading Program

Task 2　Downloading Program of Micromouse

1. Luminary driver Library

TQD-Micromouse-JD uses LM3S615 as the main control chip. The Demo program uses the Luminary DriverLib. Therefore, it requires adding the Luminary DriverLib to the installation directory of the software before downloading the program.

2. J-Link downloader

J-Link (see Fig. 1–3–2) is specially used for debugging and downloading the program of Luminary series MCU. Combined with the IAR EWARM development environment, this downloader can support the downloading and debugging of all LM3S series MCU programs.

J-Link uses USB interface to connect with computer, so it can be used freely in both desktops and laptops.

Fig. 1–3–2　J-Link

3. Connect hardware and download the program

Make sure to connect Micromouse, downloader and computer correctly before downloading the program (see Fig. 1–3–3). The top right corner of the J-Link downloader is VCC, and the top right corner of Micromouse socket is also VCC, so use a download cable to align the connection.

Fig. 1–3–3　Hardware connection

4. Click the Debug button in the IAR software to download the program

Reflection and Summary

(1) What are other common C language development softwares?

(2) IAR Embedded Workbench comes with a powerful configuration capabilities, it requires you to select the downloader model and download mode when you download the program.

020 | Micromouse Design Principles and Production Process (Intermediate)

Project 4

Basic Function Control of Micromouse

Learning objectives

(1) Understanding the human computer interaction system of Micromouse.

(2) Learning the infrared detection principle of Micromouse.

(3) Learning the motion attitude control of Micromouse.

The basic function of Micromouse is to run from the start to the destination. With limited width and a large number of turns, fast and accurate operation in a maze are inseparable from high-precision sensor detection and motor operation control. In different environments, the light intensity and the ground friction are also different, so it is necessary to use human-computer interaction to adjust the accuracy of infrared detection and the motor speeds of Micromouse.

Task 1 Human-Computer Interaction System of Micromouse

A good human-computer interaction system should transfer the decision-making of human brain to the robot quickly, and at the same time feed back the system information quickly so that the user can make a decision. In the design of TQD-Micromouse-JD, human-computer interaction system mainly refers to the circuit design of the display for debugging. Its performance directly affects the operation speed and state of Micromouse, thus affecting the results in a maze competition.

The display system of Micromouse (LED display circuit) mainly refers to the keyboard display circuit in the output circuit. It can display the coordinate in the maze of Micromouse and the wall information; the keyboard can also be set as a single step to verify each functional module, and display on the digitron, such as the speed and direction of the stepping motor. Schematic diagram of keyboard display circuit is shown in Fig. 1–4–1.

• Video

Experiment:
Digitron
Display

Chapter 1　Elementary Knowledge | 021

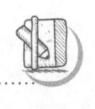

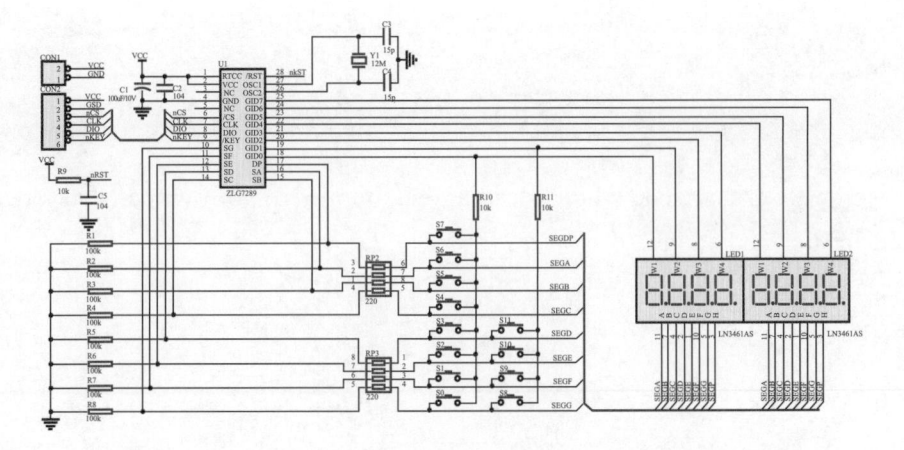

Fig. 1-4-1　Schematic diagram of keyboard display circuit

```
              Core function: Download_7289
/****************************************************************
** Function name:Download_7289
** Descriptions:Download data
** input parameters:mode=0(Download data and decode in mode 0)
**                  mode=1(Download data and decode in mode 1)
**                  mode=2(Download data without decoding)
**                  number(Digitron number, the value range is 0-7)
**                  dp=0(Decimal point is not bright)
**                  dp=1(Decimal point is bright)
**                  data(The data to be displayed)
** output parameters:None
** Returned value:None
****************************************************************/
void Download_7289 (uchar  mode,char  number,char  dp,char  data)
{
    uchar modeDat[3]={0x80,0xC8,0x90};
    uchar temp_mode;
    uchar temp_data;

    if (mode>2) {
        mode=2;
    }

    temp_mode=modeDat[mode];
    number&=0x07;
    temp_mode|=number;
    temp_data=data & 0x7F;

    if (dp==1) {
        temp_data|=0x80;
```

```
        }
        CmdDat_7289(temp_mode, temp_data);
    }
```

The flow chart:The whole idea can be simplified into two parts—system initialization and digitron display, as shown in Fig. 1-4-2.

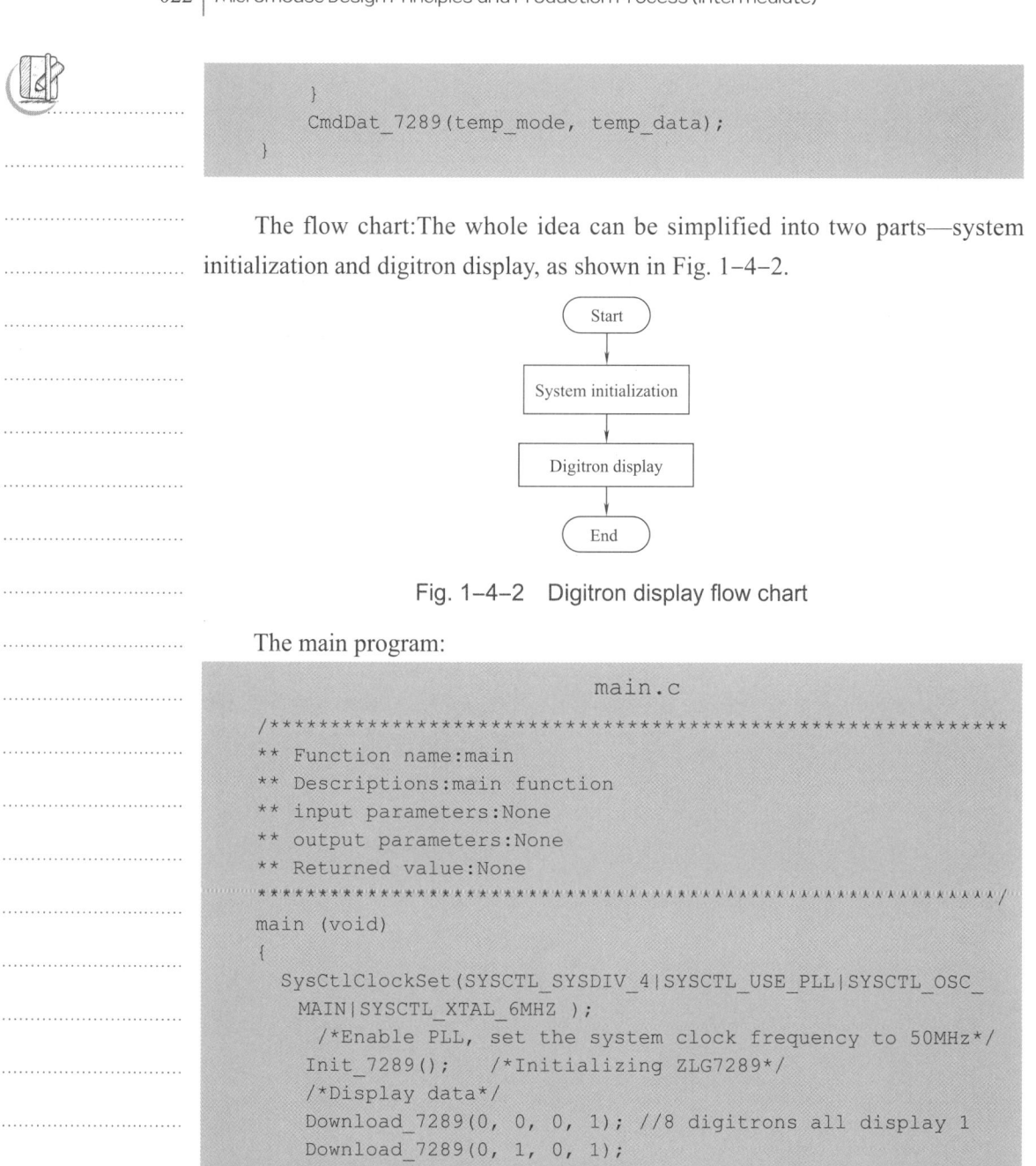

Fig. 1-4-2   Digitron display flow chart

The main program:

```
                            main.c
/***********************************************************
** Function name:main
** Descriptions:main function
** input parameters:None
** output parameters:None
** Returned value:None
***********************************************************/
main (void)
{
  SysCtlClockSet(SYSCTL_SYSDIV_4|SYSCTL_USE_PLL|SYSCTL_OSC_
    MAIN|SYSCTL_XTAL_6MHZ );
    /*Enable PLL, set the system clock frequency to 50MHz*/
    Init_7289();   /*Initializing ZLG7289*/
    /*Display data*/
    Download_7289(0, 0, 0, 1); //8 digitrons all display 1
    Download_7289(0, 1, 0, 1);
    Download_7289(0, 2, 0, 1);
    Download_7289(0, 3, 0, 1);
    Download_7289(0, 4, 0, 1);
    Download_7289(0, 5, 0, 1);
    Download_7289(0, 6, 0, 1);
    Download_7289(0, 7, 0, 1);
while (1);
}
```

Chapter 1　Elementary Knowledge ｜ 023

Task 2　Infrared Detection of Micromouse

Sensors play a very important role in the control system, and it is an indispensable part of the sensing system. There are five groups of infrared sensors on TQD-Micromouse-JD, each is composed of infrared transmitter and infrared receiver.

1. Infrared detection

Micromouse infrared detection circuit is used to detect the maze walls. There are five directions: the left, the front-left, the front, the front-right and the right. The details are as follows:

(1) Using the five groups of sensors to detect obstacles within a certain distance enables TQD-Micromouse-JD to know if there are any obstacles. It can also be used for cell identification and turning control in the process of running.

(2) The infrared sensors on left and right side can roughly judge the distance of the obstacle, and indicate three status: non-obstacle, obstacle, and near obstacle.

The principle of circuits in five directions is the same, and the detection circuit of one direction is shown in Fig. 1–4–3.

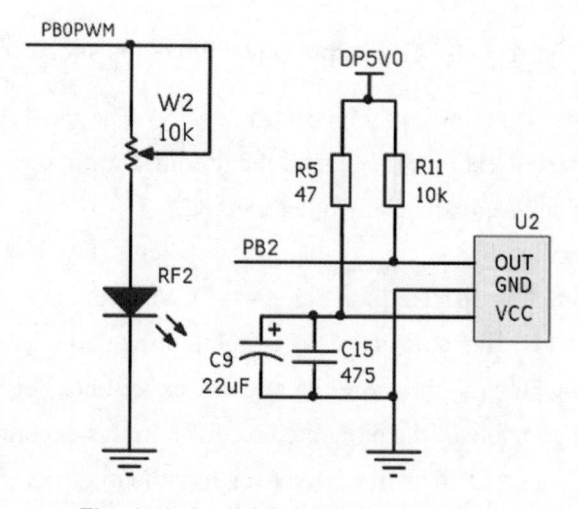

Fig. 1–4–3　Infrared detection circuit

024 | Micromouse Design Principles and Production Process (Intermediate)

RF2 is the IR transmitter, and W2 is an adjustable current-limiting resistance, which is used to adjust the intensity of the IR light.(see Fig. 1–4–4) TQD-Micromouse-JD uses IRM8601S as the IR receiver (see Fig.1-4-4), which is most sensitive to the infrared signal with carrier frequency of 38 kHz and has the farthest detection distance to it. The modulation signal of IRM8601S is a square wave with period 1,200 μs (see Fig. 1–4–5). When it detects the effective infrared signal, it will output low level, otherwise it will output high level, as shown in Fig. 1–4–6.

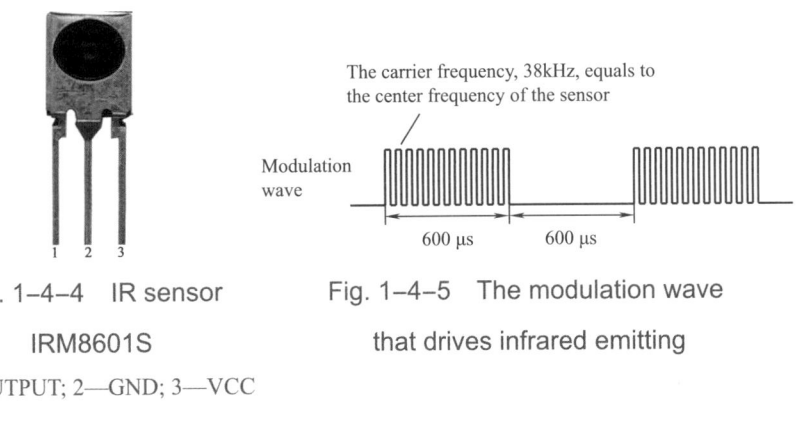

Fig. 1–4–4　IR sensor
IRM8601S
1—OUTPUT; 2—GND; 3—VCC

Fig. 1–4–5　The modulation wave
that drives infrared emitting

Fig. 1–4–6　The output waveform of the sensor

How to control the infrared emission intensity? The most direct way is to change the driving current or voltage, and the driving current could be change by adjusting resistance W2. Is there any other method?

As mentioned earlier, there is a band-pass filter in the integrated receiver. It's center frequency is 38 kHz (see Fig. 1–4–7).When the carrier frequency of infrared ray is 38 kHz, the attenuation after going through the filter is minimum. The larger the deviation is, the more the attenuation will be. This is also the key anti-interference principle of the integrated receiver. In this experiment, we adjust the detection distance of the infrared sensor by adjusting the current-limiting resistance and carrier frequency together.

• Video

Experiment:
IR range-find

Chapter 1　Elementary Knowledge | 025

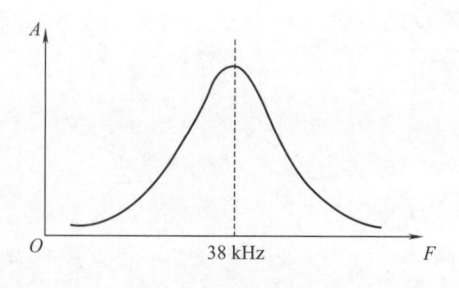

Fig. 1–4–7　Diagram of Band-pass Filtering

The IR transmitter can transmit IR light. All objects can reflect IR light with different strengths. If the distance is suitable, it can be received by the IR receiver after being reflected by the wall.

The interrupt function can also be used to read whether the infrared sensor detects the walls information. TQD-Micromouse-JD has two PWM signal generator driver modules. They are the module 1 that drives the front-right and the front-left IR sensors, and the module 2 driving the left, the front and the right IR sensors.

```
                Core function1: __irSendFreq
/*****************************************************************
** Function name:__irSendFreq
** Descriptions:Sending infrared light.
** input parameters:__uiFreq(Infrared modulation frequency)
**                  __cNumber(PWM module to be set)
** output parameters:None
** Returned value:None
*****************************************************************/
void__irSendFreq (uint__uiFreq, char__cNumber)
{
    __uiFreq=SysCtlClockGet()/__uiFreq;
    switch (__cNumber) {
    case 1:
        PWMGenPeriodSet(PWM_BASE, PWM_GEN_1, __uiFreq);
                    /*Set the cycle of PWM generator 1*/
        PWMPulseWidthSet(PWM_BASE, PWM_OUT_2, __uiFreq / 2);
                    /*Set the pulse width of PWM2 output*/
        PWMGenEnable(PWM_BASE, PWM_GEN_1);
                    /*Enable the PWM generator 1*/
        break;
    case 2:
        PWMGenPeriodSet(PWM_BASE, PWM_GEN_2, __uiFreq);
```

Micromouse Design Principles and Production Process (Intermediate)

```c
                           /*Set the cycle of PWM generator 2*/
        PWMPulseWidthSet(PWM_BASE, PWM_OUT_4, __uiFreq/2);
                  /*Set the pulse width of PWM4 output*/
        PWMGenEnable(PWM_BASE, PWM_GEN_2);
                  /*Enable the PWM generator 2*/
        break;
    default:
        break;
    }
}


                    Core function 2:__irCheck

/****************************************************************
** Function name:__irCheck
** Descriptions:Infrared detection
** input parameters:None
** output parameters:None
** Returned value:None
****************************************************************/
void__irCheck (void)
{
    static uchar ucState=0;
    static uchar ucIRCheck;
    switch (ucState) {
    case 0:
        __irSendFreq(32200, 2);
   /*Detecting the distances(short-range)of the front-right and
the front-left sides*/
        __irSendFreq(35000, 1);
   /*Detecting the distances of the front-left and the front-
right sides*/
        break;
    case 1:
        ucIRCheck=GPIOPinRead(GPIO_PORTB_BASE, 0x3e);
                       /*Reading states of IR sensors*/
        PWMGenDisable(PWM_BASE, PWM_GEN_2);
                       /*Disable PWM generator 2*/
        PWMGenDisable(PWM_BASE, PWM_GEN_1);
                       /*Disable PWM generator 1*/
        if (ucIRCheck&__RIGHTSIDE) {
            __GucDistance[__RIGHT]&=0xfd;
        } else {
            __GucDistance[__RIGHT]|=0x02;
        }
        if (ucIRCheck&__LEFTSIDE) {
            __GucDistance[__LEFT]&=0xfd;
        } else {
```

```
            __GucDistance[__LEFT]|=0x02;
        }
        if (ucIRCheck&__FRONTSIDE_R) {
            __GucDistance[__FRONTR]=0x00;
        } else {
            __GucDistance[__FRONTR]=0x01;
        }
        if (ucIRCheck&__FRONTSIDE_L) {
            __GucDistance[__FRONTL]=0x00;
        } else {
            __GucDistance[__FRONTL]=0x01;
        }
        break;
    case 2:
        __irSendFreq(36000, 2);
/*Detecting the distances(long-range)of the left,the right
and the front sides*/
        break;
    case 3:
        ucIRCheck=GPIOPinRead(GPIO_PORTB_BASE, 0x2a);
                    /*Reading states of IR sensors*/
        PWMGenDisable(PWM_BASE, PWM_GEN_2);
                    /*Disable PWM generator 2*/
        break;
    case 4:
        __irSendFreq(36000, 2);
                    /*Duplicate detection*/
        break;
    case 5:
        ucIRCheck&=GPIOPinRead(GPIO_PORTB_BASE, 0x2a);
                    /*Reading states of IR sensors*/
        PWMGenDisable(PWM_BASE, PWM_GEN_2);
                    /*Disable PWM generator 2*/
        if (ucIRCheck&__RIGHTSIDE) {
            __GucDistance[__RIGHT]&=0xfe;
        } else {
            __GucDistance[__RIGHT]|=0x01;
        }
        if (ucIRCheck&__LEFTSIDE) {
            __GucDistance[__LEFT]&=0xfe;
        } else {
            __GucDistance[__LEFT]|=0x01;
        }
        if (ucIRCheck & __FRONTSIDE) {
            __GucDistance[__FRONT]&=0xfe;
        } else {
            __GucDistance[__FRONT]|=0x01;
```

```
        }
        break;
    default:
        break;
    }
    ucState=(ucState+1)%6;
                                            /*Cycle detection*/

}
```

The flow chart: According to the characteristics of the sensor, when the frequency is closer to 38 kHz, detection distance is farther. Therefore, two different frequencies are used to drive the transmitter, such as 32.2 kHz and 36 kHz. Sensor detection flow chart is shown in Fig.1–4–8.

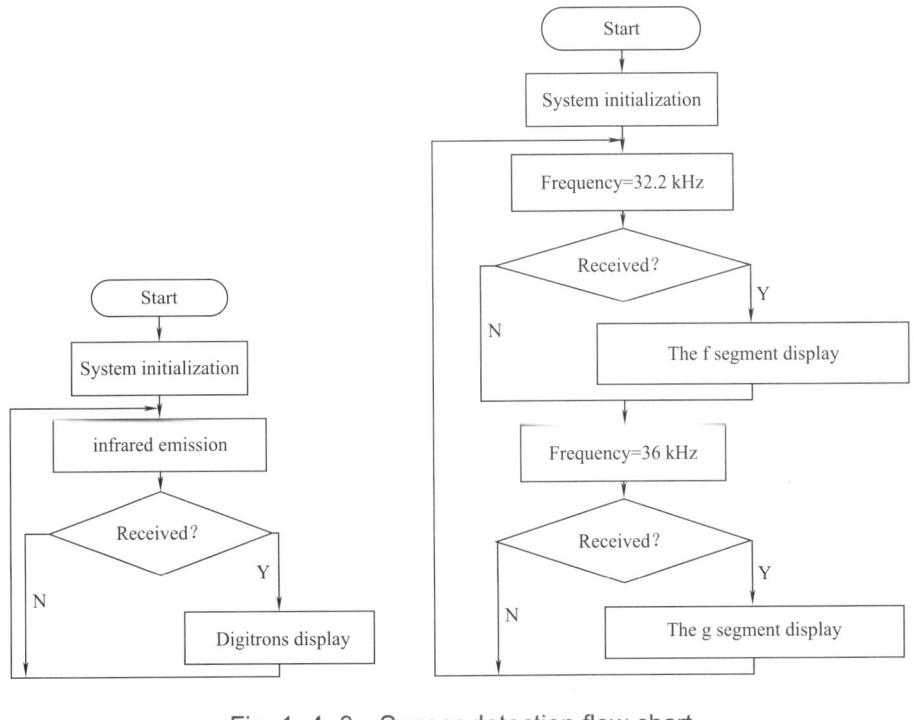

Fig. 1–4–8   Sensor detection flow chart

The main program

```
                            main. c
main (void)
{
    SysCtlClockSet(SYSCTL_SYSDIV_4|SYSCTL_USE_PLL|SYSCTL_OSC_
    MAIN|SYSCTL_XTAL_6MHZ);
```

Chapter 1　Elementary Knowledge | 029

```
/*Enable PLL, set the system clock frequency to 50 MHz*/
    IRInit();                   /*Sensor initialization*/
    SysTimerInit();             /*CLK initialization*/
    Init_7289();                /*7289 initialization*/
    while(1);                   /*Waiting for interrupt*/
}
```

2. Infrared Adjustment

(1) Debugging of the front-right and the front-left sensors. Place Micromouse in the middle of the maze passageway and adjust the second and fourth potentiometers until the g segment of the corresponding digitrons just start flashing.

(2) Debugging of the left sensor. Make Micromouse cling to the right wall and adjust the first potentiometer until the g segment of the corresponding digitron is completely lit, and the f segment flashes strongly.

(3) Debugging of the right sensor. Make Micromouse cling to the left wall and adjust the fifth potentiometer until the g segment of the corresponding digitron is completely lit, and the f segment flashes strongly.

(4) Debugging of the front sensor. Place Micromouse at the junction of the two cells and adjust the third potentiometer until the g segment of the corresponding digitron just starts flashing, as shown in Fig. 1-4-9.

Fig. 1-4-9　Debugging of the front-right and the front-left sensors, the left sensor, the right sensor, the front sensor

Task 3　Attitude Control System

The motor drive circuit of TQD-Micromouse-JD is shown in Fig. 1-4-10. The BA6845FS is a stepping motor driver, contains two H-bridge circuits, with

a maximum output current of 1 A. The logic input allows three output modes: forward, reverse, and stop.

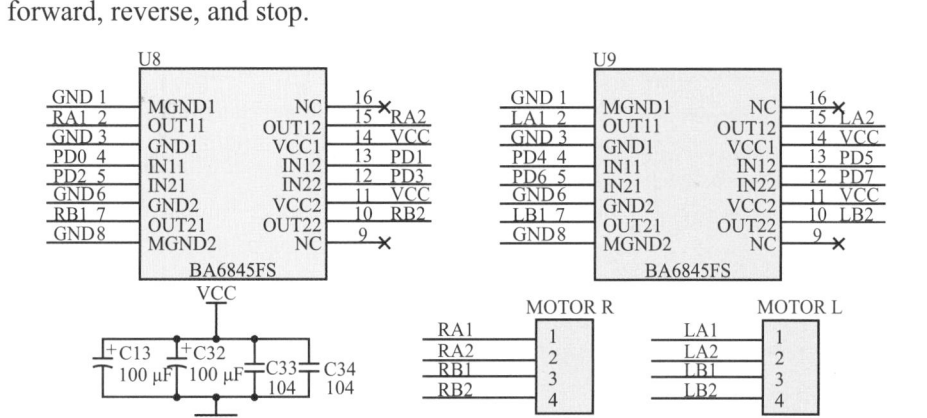

Fig. 1–4–10　Motor drive circuit

Truth table of BA6845FS is shown in Table 1–4–1.

Table 1–4–1　Truth table of BA6845FS

IN11/21	IN12/22	OUT11/21	OUT12/22	Mode
L	H	H	L	Forward
H	H	L	H	Reverse
L	L	Open	Open	Stop
H	L	Open	Open	Stop

Experiment 1　TQD-Micromouse-JD runs

After learning the motor drive, you have known that there are many parameters to consider for the motor:

(1) Motor status: Start or stop?

(2) Running direction: Forward or backward?

(3) Speed: Fast or slow?

(4) Number of steps to rotate.

(5) The number of steps that have been turned.

Therefore, a function structure can be built to store these parameters.

The struct is the same as other primitive data types (such as int, char) and the struct can be customized as needed. The function of the struct is encapsulation. The advantage of encapsulation is that it can be reused. Users do not have to care about what the struct is, as long as it is used according to the definition.

● Video

TQD-
Micromouse-
JD runs

Chapter 1　Elementary Knowledge　031

```
                    Core definition: Motor drive
/***************************************************************
Constant macro definition-Motor State
***************************************************************/
#define__MOTORSTOP          0              /*Motor stop*/
#define__WAITONESTEP        1              /*Wait one step*/
#define__MOTORRUN           2              /*Motor run*/
/***************************************************************
Constant macro definition-Motor Direction
***************************************************************/
#define__MOTORGOAHEAD       0              /*Motor go ahead*/
#define__MOTORGOBACK        1              /*Motor go back*/
/***************************************************************
Structure definition
***************************************************************/
struct__motor {
    char cState;        /*Motor state*/
    char cDir;          /*Motor running direction*/
    uint uiPulse;       /*Pulses required for motor running*/
    uint uiPulseCtr;    /*Pulses have generated by motor running*/
    int iSpeed;         /*Current speed*/
};
typedef struct__motor__MOTOR;
```

Flow chart: This experiment controls the operation of Micromouse according to its operation state, operation direction, speed, steps to be rotated and steps that have been rotated, as shown in Fig. 1–4–11.

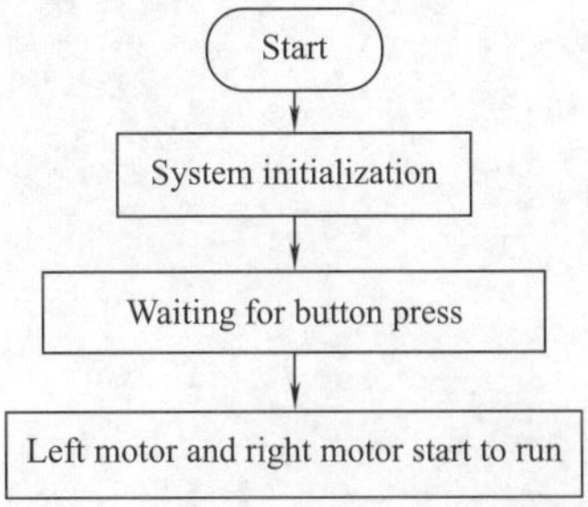

Fig. 1–4–11　The flow chart of Micromouse running

The main program:

```
                            main.c
/***************************************************************
** Function name:main
** Descriptions:main function
```

```
** input parameters:None
** output parameters:None
** Returned value:None
*****************************************************************/
main (void)
{
    SysCtlClockSet(SYSCTL_SYSDIV_4|SYSCTL_USE_PLL|SYSCTL_OSC_
      MAIN|SYSCTL_XTAL_6MHZ);
        /*Enable PLL,set the system clock frequeny to 50 MHz*/
    SysCtlPeripheralEnable( SYSCTL_PERIPH_GPIOC );
                            /*Enable GPIO C port peripherals*/
    SysCtlPeripheralEnable( SYSCTL_PERIPH_GPIOD );
                            /*Enable GPIO D port peripherals*/
    __GmRight.iSpeed=SysCtlClockGet() / 300; /*Setting the
stepping motor to rotate 300 steps per second*/
    __GmLeft.iSpeed=SysCtlClockGet() / 300; /*Setting the
stepping motor to rotate 300 steps per second*/
    GPIODirModeSet(GPIO_PORTC_BASE, KEY, GPIO_DIR_MODE_IN);
/*Set key interface as input */
    SysTickInit();                    /*CLK initialization*/
    MotorInit();                      /*Motor initialization*/
    while(1){
        if (KeyCheck()==true) {
            /*Checking whether there is a key pressed or not*/
    __GmRight.uiPulse=2000;
            /*Setting the stepping motor to rotate 2000 steps*/
    __GmRight.cDir  =FORWARD;
            /*Setting the stepping motor to rotate forward*/
    __GmRight.cState=RUN;
                /*Starting the stepping motor to rotate*/
    __GmLeft.uiPulse=2000;
            /*Setting the stepping motor to rotate 2000 steps*/
    __GmLeft.cDir  =FORWARD;
            /*Setting the stepping motor to rotate forward*/
    __GmLeft.cState=RUN;
                /*Starting the stepping motor to rotate*/
            while(__GmRight.cState!=STOP&&__GmLeft.cState!=STOP);
        }
    }
}
```

● Video

Experiment:
Differential-
speed control

Experiment Two   Differential-speed control

It is generally known that, when the speeds of the left motor and the right motor are the same, it will walk along a straight line; otherwise, it will turn according to the differential-speed. Next, we will try to control Micromouse to

Chapter 1 Elementary Knowledge | 033

turn through experiment.

Flow chart: First, set the speeds of the left motor and the right motor, and control Micromouse to turn according to the differential-speed.

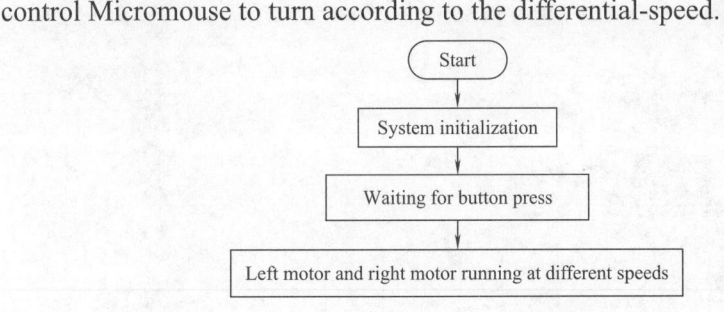

Fig. 1–4–12   Differential-speed control of Micromouse

The main program:

```
                               main.c
/***************************************************************
** Function name:main
** Descriptions:main function
** input parameters:None
** output parameters:None
** Returned value:None
****************************************************************/
main (void)
{

  SysCtlClockSet(SYSCTL_SYSDIV_4|SYSCTL_USE_PLL|SYSCTL_OSC_
  MAIN|SYSCTL_XTAL_6MHZ);
      /*Enable PLL,set the system clock frequency to 50 MHz*/

  SysCtlPeripheralEnable( SYSCTL_PERIPH_GPIOC );
                  /*Enable the peripheral clock of GPIO C. */
  SysCtlPeripheralEnable( SYSCTL_PERIPH_GPIOD );
  GPIODirModeSet(GPIO_PORTC_BASE, KEY, GPIO_DIR_MODE_IN);
                              /*Set key interface as input */
  SysTickInit();                        *CLK initialization*/
  MotorInit();                       /*Motor initialization*/
while(1)
    {
if (KeyCheck()==true)
          /*Checking whether there is a key pressed or not*/
{
__GmRight.iSpeed=SysCtlClockGet() /200;
  /*Setting the stepping motor to rotate 200 steps per second*/
__GmLeft.iSpeed=SysCtlClockGet() / 100;
```

034 | Micromouse Design Principles and Production Process (Intermediate)

```
        /*Setting the stepping motor to rotate 100 steps per second*/
__GmRight.uiPulse=1000;
            /*Setting the stepping motor to rotate 1000 steps*/
__GmRight.cDir=FORWARD;
                /*Setting the stepping motor to rotate forward*/
__GmRight.cState=RUN;
            /*Starting the stepping motor to rotate*/
__GmLeft.uiPulse=500;
              /*Setting the stepping motor to rotate 500 steps*/
__GmLeft.cDir  =FORWARD;
  /*Setting the stepping motor to rotate forward*/
__GmLeft.cState=RUN;
            /*Starting the stepping motor to rotate*/
      TimerLoadSet(TIMER0_BASE, TIMER_A, __GmRight.iSpeed);
      TimerLoadSet(TIMER1_BASE, TIMER_A, __GmLeft.iSpeed);
      while(__GmRight.cState!=STOP&&__GmLeft.cState!=STOP);
    }
  }
}
```

Experiment 3   The journey control of motor running

We have learned how to drive motor and control its speed through the above two experiments. Then how to identify the location of Micromouse in the maze? We change both __GmRight.uiPulse and __GmLeft.uiPulse in "Micromouse running" to 500 and watch the distance ran by Micromouse. And then we change it to 750 and observe. Through the experiment, we can draw to the conclusion that when the wheel hub is settled, a Micromouse can run 4 cells with 500 steps, and 6 cells with 750 steps. Micromouse calculates the number of cells ran by recording the steps ran and combining with the diameter of wheel hub and the steps needed for each cell.

When TQD-Micromouse-JD uses the standard wheel hub, for each cell, it need to run 125 steps.

```
                Core function1: mouseGoahead
/********************************************************************
** Function name:mouseGoahead
** Descriptions:Going ahead Ncells
** input parameters:iNblock(cells need to be done)
** output parameters:None
** Returned value:None
********************************************************************/
```

```c
void mouseGoahead (char  cNBlock)
{
    char cL=0, cR=0, cCoor=1;
    if (__GmLeft.cState)              //Judging the motor state
    {
        cCoor=0;
    }
    __GucMouseState=__GOAHEAD;
    __GiMaxSpeed=MAXSPEED;
    __GmRight.cDir=__MOTORGOAHEAD;
    __GmLeft.cDir=__MOTORGOAHEAD;
    __GmRight.uiPulse=__GmRight.uiPulse+cNBlock*ONEBLOCK-6;
    __GmLeft.uiPulse=__GmLeft.uiPulse+cNBlock*ONEBLOCK-6;
    __GmRight.cState=__MOTORRUN;
    __GmLeft.cState=__MOTORRUN;

    while (__GmLeft.cState!=__MOTORSTOP) {
        if (__GmLeft.uiPulseCtr>=ONEBLOCK) {
                        /*Judge whether complete one cell*/
            __GmLeft.uiPulse-=ONEBLOCK;
            __GmLeft.uiPulseCtr-=ONEBLOCK;
            if (cCoor)
            {
                cNBlock--;
                __mouseCoorUpdate();  /*Coordinate update*/
            }
            else
            {
                cCoor = 1;
            }
        }
        if (__GmRight.uiPulseCtr>=ONEBLOCK) {
                /*Judging whether complete one cell or not*/
            __GmRight.uiPulse-=ONEBLOCK;
            __GmRight.uiPulseCtr-=ONEBLOCK;
        }
        if (__GucDistance[__FRONT]) {
        /*If there is a wall in front, jump out of the program.*/
            __GmRight.uiPulse=__GmRight.uiPulseCtr+70;
            __GmLeft.uiPulse=__GmLeft.uiPulseCtr +70;
            while (__GucDistance[ __FRONT]) {
                if ((__GmLeft.uiPulseCtr+20)>__GmLeft.uiPulse) {
                    goto End;
                }
            }
        }
        if (cNBlock<2) {
```

036 | Micromouse Design Principles and Production Process (Intermediate)

```
                        if (cL) {                        /*Judging whether the
left side is allowed to be detected or not. */
          if ((__GucDistance[ __LEFT] & 0x01)==0) { /*If there is
a branch on the left, jump out of the program*/
                        __GmRight.uiPulse=__GmRight.uiPulseCtr+74;
                        __GmLeft.uiPulse=__GmLeft.uiPulseCtr +74;
                        while ((__GucDistance[ __LEFT] & 0x01)==0) {
                            if ((__GmLeft.uiPulseCtr+20)>__GmLeft.
                            uiPulse) {
                                goto End;
                            }
                        }
                    }
                } else {
        if ( __GucDistance[ __LEFT] & 0x01) { /*When there is a wall
on the right, it is allowed to detect the left side*/
                        cL=1;
                    }
                }
                if (cR) {        *Judging whether the right side is
allowed to be detected or not */
          if ((__GucDistance[ __RIGHT]&0x01)==0) { /*If there is a
branch on the right, jump out of the program*/
                        __GmRight.uiPulse=__GmRight.uiPulseCtr+74;
                        __GmLeft.uiPulse=__GmLeft.uiPulseCtr +74;
                        while ((__GucDistance[ __RIGHT] & 0x01)==0) {
                            if ((__GmLeft.uiPulseCtr+20)>__GmLeft.uiPulse){
                                goto End;
                            }
                        }
                    }
                } else {
        if ( __GucDistance[ __RIGHT] & 0x01) { /*When there is a wall on
the left, it is allowed to detect the right side*/
                    }
                }
            }
        }

    End:__mouseCoorUpdate();
    }
                Core function2: __mouseCoorUpdate
    /*****************************************************
    ** Function name:__mouseCoorUpdate
    ** Descriptions:Updating the coordinates according to the
number of running steps
    ** input parameters:None
    ** output parameters:None
```

```
** Returned value:None
****************************************************************/
void__mouseCoorUpdate (void)
{

  if(GucMouseDir==0)
  {
     GmcMouse.cY++;
  }
   Download_7289(1, 3, 0, GmcMouse.cY%10);
}
```

Flow chart: This experiment calculate the current cell number by recording the number of steps ran, and show through 7289, as shown in Fig. 1–4–13.

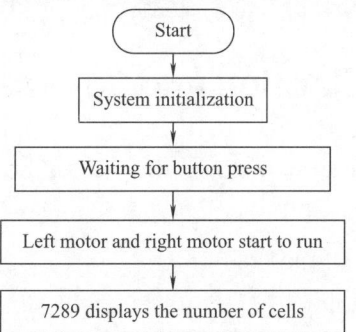

Fig. 1–4–13   The flow chart of journey control

The main program:

```
                        main.c
   /****************************************************************
   ** Function name:main
   ** Descriptions:main function
   ** input parameters:None
   ** output parameters:None
   ** Returned value:None
   ****************************************************************/
   main (void)
   {
      uchar ucRoadStat=0;        /*Counting the number of
branches that can go ahead in a certain coordinate*/
      mouseInit();   /*Initialization of the bottom driver program*/
      Init_7289();    /*Initialization of the display module*/
      while (1) {
      switch (GucMouseTask) {   /*Processing of state machine*/
        case WAIT:
```

```
                sensorDebug();
                voltageDetect();
                delay(100000);
                if (keyCheck()==true) {
                                    /*Waiting for button press*/
                    Reset_7289();/*Reset 7289*/
                    GucMouseTask=START;
                }
                break;
            case START:
             mouseGoahead(5);      //6、7、8
             while(1);
            default:
                break;
            }
        }
    }
```

Reflection and Summary

(1) How does Micromouse realize human-computer interaction?

(2) What is the function of IR sensors? How to improve the detection accuracy?

(3) How many types of motors?

(4) The PWM technology is often used for motor speed regulation, which has the advantages of fast response, high precision, etc.

Chapter **2**

Comprehensive Practice

This chapter mainly introduces the debugging methods of forward-moving and precise turning, and introduces the basic principles and operations of Micromouse innovation competition. Students can master the turning function and the application of various turning combination faster. According to the requirements in the task papers randomly selected by the judges, the competitors should complete the programming and realize the corresponding functions on site.

Micromouse Design Principles and Production Process (Intermediate)

Chapter 2　Comprehensive Practice　041

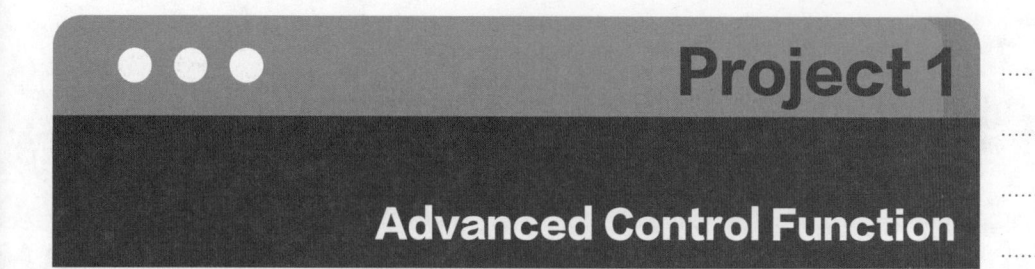

Project 1

Advanced Control Function

 Learning objectives

(1) Master the principle and control method of forward-moving.

(2) Learn to control the accurate turning of Micromouse.

The operation of Micromouse in a maze can be simplified into two parts: straight movement and turning. Straight movement refers to Micromouse in the maze through the detection of two side walls to correct it's attitude and avoid touching. Turning refers to the intelligent mouse's precise turning of 180° and 90°, combined with straight movement to the final destination.

Task 1　Straight Movement Control

The sensors on Micromouse are its "eyes", and its two wheels are its "feet". If you want it to move and work as required, you need to learn to control the use of its "eyes" and "feet". When people are moving, they observe the surrounding environment through their eyes, and use these information as the basis to command the foot movement. This is also the case with Micromouse, which use sensors to obtain external information and use these as a basis to control its wheels movement.

When Micromouse is left offset, increase the left motor speed or decrease the right motor speed.

When Micromouse is right offset, increase the right motor speed or decrease the left motor speed.

The method of "waiting one step" is adopted here: When Micromouse is offset, the outer motor waiting one step, so as to achieve the purpose of correction.

Video

Experiment:
Attitude
correction

```
                    Core function1: Timer0A_ISR
/**************************************************************
** Function name:Timer0A_ISR
** Descriptions:Timer0 interrupt handler
** input parameters:None
** Output parameters:None
** Returned value:None
**************************************************************/
void Timer0A_ISR(void)
{
    static char n=0,m=0;
    TimerIntClear(TIMER0_BASE, TIMER_TIMA_TIMEOUT);
                            /*Clearing Timer0 interrupt*/
    switch (__GmRight.cState) {
    case__MOTORSTOP:              /* Motor stop, clear speed and
pulse value at the same time*/
            __GmRight.iSpeed=0;
            __GmRight.uiPulse=0;
            __GmRight.uiPulseCtr=0;
        break;
    case__WAITONESTEP:
            __GmRight.cState=__MOTORRUN;
        break;
    case__MOTORRUN:
            if (__GucMouseState==__GOAHEAD) {
            /* Fine tune motor speed according to sensor states*/
            if (__GucDistance[__FRONTL]&&(__GucDistance[__FRONT]==0)) {
                    if (n==1) {
                        __GmRight.cState=__WAITONESTEP;
                    }
                    n++;
                    n%=2;   /*Running one step and waiting one
step are equivalent to halving the speed*/
                } else {
                    n=0;
                }

            if ((__GucDistance[__RIGHT]==1)&&(__GucDistance[__LEFT]==0))
                {
                    if(m==1)
                    {
                        __GmRight.cState=__WAITONESTEP;
                    }
                    m++;
                    m%=2;
                } else
```

```
                    {
                         m=0;
                    }
               }
          __rightMotorContr();                    /*Motor driver*/
          break;

     default:
          break;
     }
     if (__GmRight.cState !=__MOTORSTOP) {
                    /*Judging whether the task is completed or not*/
        __GmRight.uiPulseCtr++;                 /*Step count*/
 __speedContrR();                            /*Speed regulation*/
          if (__GmRight.uiPulseCtr>=__GmRight.uiPulse) {
               __GmRight.cState=__MOTORSTOP;
               __GmRight.uiPulseCtr=0;
               __GmRight.uiPulse=0;
               __GmRight.iSpeed=0;
          }
     }
}
               Core function2: Timer1A_ISR
/*****************************************************************
** Function name:Timer1A_ISR
** Descriptions:Timer1 interrupt handler
** input parameters:__GmLeft.cState(Time sequence status of
                    driving stepping motor)
**                  __GmLeft.cDir(Motor running direction)
** output parameters:None
** Returned value:None
*****************************************************************/
void Timer1A_ISR(void)
{
     static char n=0, m=0;
     TimerIntClear(TIMER1_BASE, TIMER_TIMA_TIMEOUT);
                         /*Clearing Timer1 interrupt*/
     switch (__GmLeft.cState) {
     case __MOTORSTOP: /*Motor stop, clear speed and pulse
value at the same time*/
          __GmLeft.iSpeed=0;
          __GmLeft.uiPulse=0;
          __GmLeft.uiPulseCtr=0;
          break;
     case __WAITONESTEP:                      /*Waiting one step*/
          __GmLeft.cState=__MOTORRUN;
```

```c
            break;
    case __MOTORRUN:                          /*Motor runs*/
        if (__GucMouseState==__GOAHEAD) {
          /*Fine tune motor speed according to sensor states*/
          if (__GucDistance[__FRONTR]&&(__GucDistance[__FRONT]==0)) {
                if (n==1) {
                    __GmLeft.cState=__WAITONESTEP;
                }
                n++;
                n%=2;
            } else {
                n=0;
            }
    if ((__GucDistance[__LEFT]==1)&&(__GucDistance[__RIGHT]==0)) {
                if(m==1) {
                    __GmLeft.cState=__WAITONESTEP;
                }
                m++;
                m%=2;
            } else {
                m=0;
            }
        }
        __leftMotorContr();                        /*Motor driver*/
        break;

    default:
        break;
    }
    if (__GmLeft.cState !=__MOTORSTOP) {
            /*Judging whether the task is completed or not*/
        __GmLeft.uiPulseCtr++;               /*Step count*/
        __speedContrL();                     /*Speed regulation*/
        if (__GmLeft.uiPulseCtr>=__GmLeft.uiPulse) {
            __GmLeft.cState=__MOTORSTOP;
            __GmLeft.uiPulseCtr=0;
            __GmLeft.uiPulse=0;
            __GmLeft.iSpeed=0;
        }
    }
}
```

Flow chart is shown as Fig. 2–1–1.

Chapter 2   Comprehensive Practice | 045

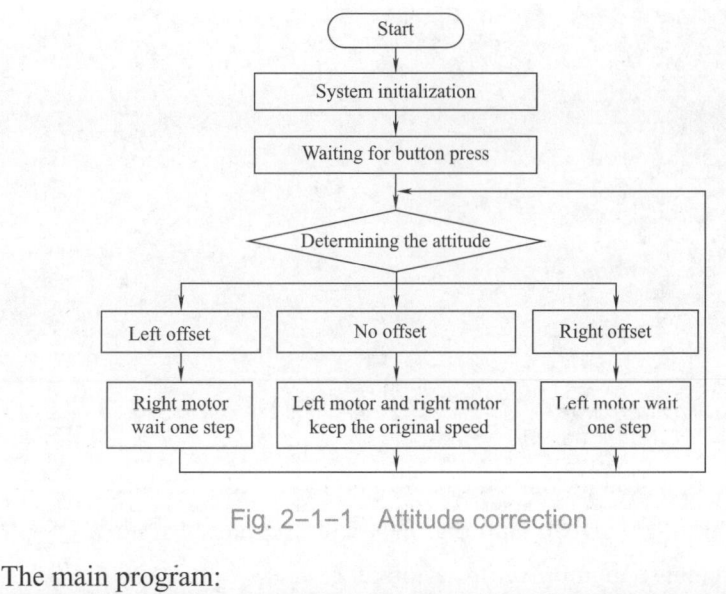

Fig. 2-1-1   Attitude correction

The main program:

```
                          main.c
/**************************************************************
** Function name:main
** Descriptions:main function
** input parameters:None
** output parameters:None
** Returned value:None
**************************************************************/
main (void)
{
uchar n=0;    /*The number of coordinates with multiple branches*/
uchar ucRoadStat=0;
       /*Counting the number of directions can move forward*/
uchar ucTemp=0;    /*Used for coordinate conversion in START state*/
mouseInit();         /*Underlying driver initialization*/
Init_7289();         /*Initialization of the display module*/
    while (1) {
        switch (GucMouseTask) { /*Processing of state machine*/
        case WAIT:
            sensorDebug();
            voltageDetect();
            delay(100000);
            if (keyCheck()==true) {  /*Waiting for button press*/
                Reset_7289();         /*Reset ZLG7289*/
                GucMouseTask=START;
            }
            break;
```

```
                case START:
                        crosswayChoice();  /*Choosing the forward
direction with the right-hand rule*/
                        mazeSearch();  /*Moving forward one cell*/
                    break;
                default:
                    break;
                }
            }
        }
```

Task 2   Accurate Turning Control

In addition to straight movement, Micromouse will encounter a lot of turns, such as 90° turning and 180° turning. Fast and accurate turning is an effective way for Micromouse to improve its performance.

```
                Core function1:mouseTurnright
    /***********************************************************
    ** Function name:mouseTurnright
    ** Descriptions:Turn right
    ** input parameters:None
    ** output parameters:None
    ** Returned value:None
    ***********************************************************/
    void mouseTurnright(void)
    {
        while (__GmLeft.cState!=__MOTORSTOP);
        while (__GmRight.cState!=__MOTORSTOP);
        __GucMouseState=__TURNRIGHT;   /*Starting to turn right*/
        __GmRight.cDir=__MOTORGOBACK;
        __GmRight.uiPulse=45;
        __GmLeft.cDir=__MOTORGOAHEAD;
        __GmLeft.uiPulse=45;
        __GmRight.cState=__MOTORRUN;
        __GmLeft.cState=__MOTORRUN;
        GucMouseDir=(GucMouseDir+1)%4;              /*Direction mark*/
        while (__GmLeft.cState!=__MOTORSTOP);
        while (__GmRight.cState!=__MOTORSTOP);
        __mazeInfDebug();
        __delay(100000);
    }
                Core function2: mouseTurnleft
    /***********************************************************
```

Text

Accurate
turning
explanation

```c
** Function name:mouseTurnleft
** Descriptions:Turn left
** input parameters:None
** output parameters:None
** Returned value:None
*****************************************************************/
void mouseTurnleft(void)
{
    while (__GmLeft.cState!=__MOTORSTOP);
    while (__GmRight.cState!=__MOTORSTOP);
    __GucMouseState=__TURNLEFT;
    __GmRight.cDir=__MOTORGOAHEAD;/*Starting to turn left*/
    __GmRight.uiPulse=47;
    __GmLeft.cDir=__MOTORGOBACK;
    __GmLeft.uiPulse=47;
    __GmRight.cState=__MOTORRUN;
    __GmLeft.cState=__MOTORRUN;
    GucMouseDir=(GucMouseDir+3)%4;              /*Direction mark*/
    while (__GmLeft.cState!=__MOTORSTOP);
    while (__GmRight.cState!=__MOTORSTOP);
    __mazeInfDebug();
    __delay(100000);
}
```

Core function 3: mouseTurnback

```c
/****************************************************************
** Function name:mouseTurnback
** Descriptions:Turn back
** input parameters:None
** output parameters:None
** Returned value:None
*****************************************************************/
void mouseTurnback(void)
{
    while (__GmLeft.cState!=__MOTORSTOP); /*Waiting to stop*/
    while (__GmRight.cState!=__MOTORSTOP);
    __GucMouseState=__TURNBACK;       /*Starting to turn back*/
    __GmRight.cDir=__MOTORGOBACK;
    __GmRight.uiPulse=90;//162*10;
    __GmLeft.cDir=__MOTORGOAHEAD;
    __GmLeft.uiPulse=90;//162 * 10;
    __GmLeft.cState=__MOTORRUN;
    __GmRight.cState=__MOTORRUN;
    GucMouseDir=(GucMouseDir+2)%4;              /*Direction mark*/
    while(__GmLeft.cState!=__MOTORSTOP);
    while(__GmRight.cState!=__MOTORSTOP);
    __mazeInfDebug();
```

```
    __delay(100000);
}
```

Flowchart: We take right turn as an example to realize the accurate turning control of Micromouse, as shown in Fig. 2–1–2.

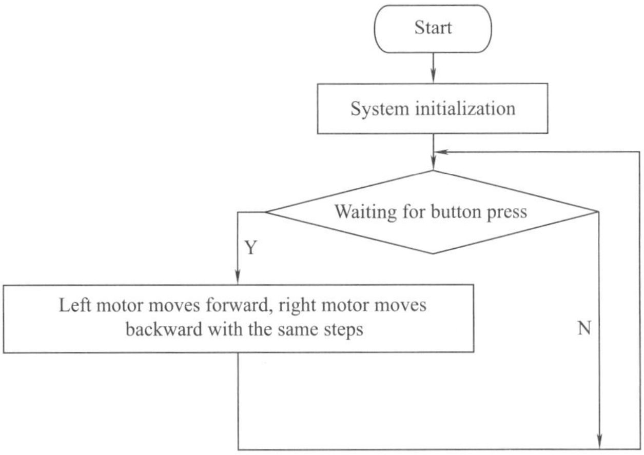

Fig. 2–1–2   Accurate turning control

The main program:

```
                              main.c
/****************************************************************
** Function name:main
** Descriptions:main function
** input parameters:None
** output parameters:None
** Returned value:None
****************************************************************/
main (void)
{
    uchar n=0;  /*The number of coordinates with multiple branches*/
    uchar ucRoadStat=0;      /*Counting the number of branches
that can go ahead in a certain coordinate*/
    uchar ucTemp=0;  /*Used for coordinate conversion in start state*/
    mouseInit();    /*Initialization of the bottom driver program*/
    Init_7289();   /*Initialization of the display module*/
while (1) {
        switch (GucMouseTask) {  /*Processing of state machine*/
            case WAIT:
            sensorDebug();
            voltageDetect();
            delay(100000);
```

```
            if (keyCheck()==true) {    /*Waiting for button press*/
            Reset_7289();              /*Reset ZLG7289*/
            GucMouseTask = START;
            }
            break;
        case START:
            mouseTurnright();                  /*Turn right*/
            //mouseTurnback();
            //mouseTurnleft();
            break;
        default:
            break;
        }
    }
}
```

Reflection and Summary

(1) How does Micromouse turn?

(2) What is the difference between counter-clockwise and clockwise of 180° turning?

(3) In order to avoid mutual interference, five groups of infrared sensors adopt the way of alternating emission. High frequency detection of long-distance obstacles, low frequency detection of short-distance obstacles, multiple groups of data together to judge the information of obstacles, so as to achieve high-precision control of speed and position of Micromouse.

050　Micromouse Design Principles and Production Process (Intermediate)

Project 2

Actual Combat Tasks

Learning objectives

(1) Master the basic functions of Micromouse.

(2) Learning the applications of Micromouse IR sensors, interrupt functions and motor-drive.

(3) Learn how to use and control the turning function of Micromouse.

(4) Learn the realization method of multiple turn combination application of Micromouse.

In a Micromouse competition, in order to investigate the real level of competitors, except the speed link, there will be field programming projects. According to the task given by the referee at the scene, write the program and realize the corresponding functions.

Task 1　Using of Debugging Board 7289

1. Task description

This task exams students' familiarity with programming environment and knowledge of digitron display function. Programming and realizing 7289 display specified data, as shown in Fig. 2-2-1.

Fig. 2-2-1　7289 display specified data

2. Task objectives

Learning of 7289 display unit and display any data.

3. Task content

Programming and realizing digitrons display 0,1,2,…,7 in turn.

Chapter 2　Comprehensive Practice | 051

4. Application principle

In the task of human-computer interaction system, the application of 7289 software package has been studied. This task attempts to modify the data to be displayed.

5. Task flow chart

The whole idea of the experiment is simplified into two parts: system initialization and digitron display, as shown in Fig. 2-2-2.

Fig. 2-2-2　Digitrons display designated data

6. Specific steps

Step 1: Open the IAR EWARM development environment and create a new project.

Step 2: Adding the header files into the project according to the software usage method.

Step 3: Downloading the program to "Micromouse".

Step 4: Disconnecting Micromouse from the computer, turn on the power, run the program, and observe the display of digitrons.

7. Application program

```
                        main.C
    /****************************************************************
    ** Function name:main
    ** Descriptions:main function
    ** input parameters:None
    ** output parameters:None
    ** Returned value:None
    ****************************************************************/
    main (void)
    {
    SysCtlClockSet( SYSCTL_SYSDIV_4| SYSCTL_USE_PLL | SYSCTL_OSC_
MAIN |SYSCTL_XTAL_6MHZ );              /*Enable PLL, 50MHz*/
        Init_7289();                   /*ZLG7289 initialization*/
        Download_7289(0, 0, 0, 0);   /*No.0 digitron displays 0*/
        Download_7289(0, 1, 0, 1);   /*No.1 digitron displays 1*/
        Download_7289(0, 2, 0, 2);   /*No.2 digitron displays 2*/
        Download_7289(0, 3, 0, 3);   /*No.3 digitron displays 3*/
```

```
        Download_7289(0, 4, 0, 4);    /*No.4 digitron displays 4*/
        Download_7289(0, 5, 0, 5);    /*No.5 digitron displays 5*/
        Download_7289(0, 6, 0, 6);    /*No.6 digitron displays 6*/
        Download_7289(0, 7, 0, 7);    /*No.7 digitron displays 7*/
    while (1);
    }
```

Task 2 Non-Contact Start/Stop Control

1. Task description

This task exams students' mastery of knowledge of infrared function, interrupt function, and motor drive.

Programming and realizing a non-contact method (no other buttons other than the power switch) to start Micromouse. Micromouse shall not touch any walls during the whole moving process. After running for a distance, Micromouse must be stopped in the same non-contact way. Two parking positions are shown in Fig. 2–2–3.

Fig. 2–2–3 Non-contact start/stop control

2. Task objectives

Learn to start and stop Micromouse in a non-contact way.

3. Task content

Programming and realizing the following operations: start Micromouse when blocking the front sensor; and stop it when there are walls on three sides.

4. Application principle

Infrared detection of obstacle information can be used as the basis for Micromouse to start, and being surrounded with three walls can be used as a judgment condition for Micromouse to stop.

5. Task flow chart

The flow chart of non-contact start/stop control is shown in Fig. 2–2–4.

● Video

Experiment:
Non-contact
start/stop
control

Chapter 2 Comprehensive Practice | 053

```
                    Start
                      │
                      ▼
            System initialization
                      │
                      ▼
            Reading sensor data ◄──────────┐
                      │                     │
                      ▼                     │
                 Obstacle?  ──N─────────────┤
                  │                         │
                  Y                         │
                  ▼                         │
            Micromouse start                │
                  │                         │
                  ▼                         │
          Walls on three sides?             │
             │              │               │
             Y              N               │
             ▼              ▼               │
           Stop          Continue           │
             │              │               │
             └──────────────┴───────────────┘
```

Fig. 2-2-4   Non-contact start/stop control flow chart

6. Specific steps

Step 1: Open the IAR EWARM development environment and create a new project.

Step 2: Adding the header files into the project according to the software usage method.

Step 3: Downloading the program to "Micromouse".

Step 4: Disconnecting Micromouse from the computer, turn on the power, run the program. Micromouse starts when blocking the front sensor with hand, and it stops when there are walls on three sides.

7. Application program

```
          Core function: Modified mouseTurnback
/*****************************************************************
** Function name:mouseTurnback
** Descriptions:Turn back
** input parameters:None
** output parameters:None
** Returned value:None
*****************************************************************/
void mouseTurnback(void)
{
    while (__GmLeft.cState!=__MOTORSTOP);   /*Waiting to stop*/
```

```
        while (__GmRight.cState!=__MOTORSTOP);
        /*
    Waiting for the button press, otherwise it will be stop all the
time
        */
        while (1) {
                if (keyCheck()==true) {
                    break;
                }
            }
        .
        .
        .
```

Task 3   Movement Control in Picture-8-Shaped Path

1. Task description

This task exams students' combined use of different kinds of turns.

Programming and realizing the running trajectory of Micromouse resembling picture-8, as shown in Fig. 2-2-5. Micromouse shall not touch any wall during the whole moving process and is required to return to where it starts in the end. (The start direction can be either left or right).

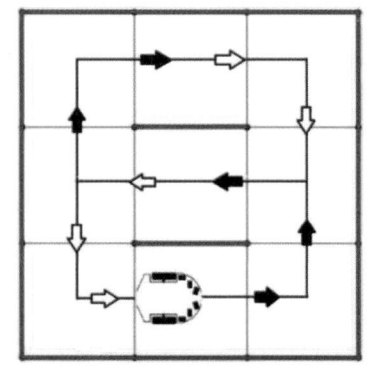

Fig. 2-2-5   Micromouse running trajectory

2. Task objectives

Learn how to make different kinds of turns.

3. Task content

Programming and realizing the picture-8-shaped trajectory of Mcromouse.

4. Application principle

Through the learning above, we understand that Micromouse can make turns according to its own algorithm while running. Through the analysis of Fig. 2-2-5, we

Chapter 2 Comprehensive Practice | 055

understand that the turning direction of Micromouse at multiple crossings either left or right. Therefore, different turning functions should be adopted according to the number of turns.

5. Task flow chart

The general idea of this task is simplified as the adopting of the turning functions. When there is only one crossing, extra judgment is unnecessary, so students needn't take it into consideration, as shown in Fig. 2–2–6.

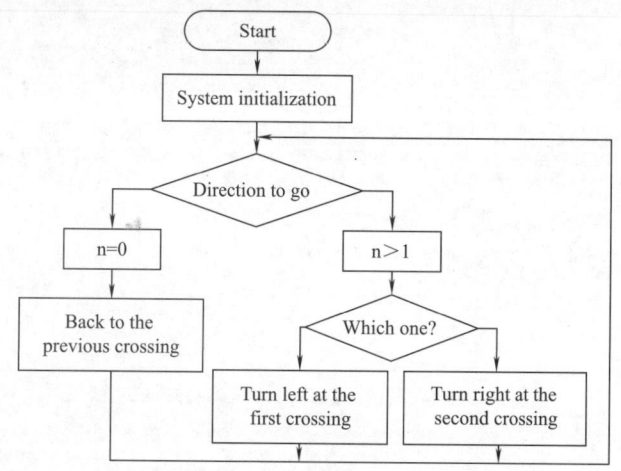

Fig. 2–2–6   The flow chart of Picture-8-shaped trajectory

6. Specific steps

Step 1: Open the IAR EWARM development environment and create a new project.

Step 2: According to the software manual, add the header files to the project.

Step 3: Downloading the program to "Micromouse".

Step 4: Disconnecting Micromouse from the computer, turn on the power, and run the program.

7. Application program

```
              Core function: crosswayChoice
/********************************************************
** Function name:crosswayChoice
** Descriptions:selecting a branch as the forward direction
********************************************************/
void crosswayChoice (void)
```

```
{
    switch (SEARCHMETHOD) { //Judge the turning direction according
to SEARCHMETHOD
        case RIGHTMETHOD:
            mouseTurnright();              //Turn right
            break;
        case LEFTMETHOD:
            mouseTurnleft();               //Turn left
            break;
        case CENTRALMETHOD:
            centralMethod();
            break;
        case FRONTRIGHTMETHOD:
            frontRightMethod();
            break;
        default:
        break;
        }
    }
```

Reflection and Summary

(1) Are there any other non-contact methods available for starting and stopping Micromouse?

(2) Micromouse adopts the two-wheel difference speed turning method. The greater the speed difference is, the greater the number of steps is, the greater the turning angle; the smaller the speed difference is, the smaller the number of steps is, the smaller the turning angle.

Chapter 3

Advanced Skills and Competitions

The software and hardware of Micromouse and the basic programming and debugging methods have been introduced previously. This chapter mainly introduces the optimization algorithm according to the requirements of the IEEE International Standard Micromouse Competition. Mastering the specifications of competition enables the participants to complete the maze search and the best path selection at the fastest speed. Analyzing the key points of Micromouse competition cases, so as to prepare for participating in IEEE International Standard Micromouse competition.

Micromouse Design Principles and Production Process (Intermediate)

Chapter 3　Advanced Skills and Competitions　059

Project 1

Path Planning and Decision Algorithm

Learning objectives

(1) Learning path planning and decision algorithm of Micromouse.

(2) Learn the principle of Micromouse path planning.

How can Micromouse runs quickly in a maze? The main task of Micromouse is to complete the maze search and optimal path selection according to IEEE International Standard rules. It is a competition that exams the detection, analysis and decision-making ability of a system to an unknown environment. Here is a brief introduction.

Task 1　Common Strategies of Maze Searching

Maze search method: Without knowing the maze path, Micromouse must first explore all the cells in the maze until it reaches the destination. Micromouse going this procedure must know its position and posture at any time, and record whether there are walls around the cells. In order to save time during this process, we should try our best to avoid repeatedly searching the places.

Then how to explore the maze? Usually, there are two strategies: (1) Reaching the destination as soon as possible. (2) Searching the whole maze.

Both strategies have advantages and disadvantages. The first strategy can shorten the time needed to explore the maze, but we might not be able to obtain the data of the entire maze. If the path found is not the optimal path of the maze, which will affect the final sprint time of Micromouse. Using the second strategy can get the information of the entire maze so that we can find the optimal path. But, this method will take a long time to search.

Task 2　The Basic Rules of Maze Searching

Video

The basic rules of maze searching (right hand)

Video

The basic rules of maze searching (left hand)

Video

The basic rules of maze searching (central)

There are three common rules for searching: the right-hand rule, the left-hand rule and the central rule, as shown in Fig. 3–1–1.

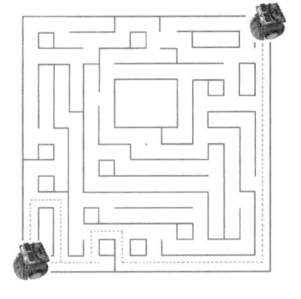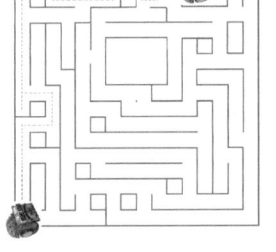

(a) The right-hand rule　　　　　(b) The left-hand rule　　　　　(c) The central rule

Fig. 3–1–1　The right-hand rule,the left-hand rule, the central rule

The right-hand rule: when there are multiple choices of heading, the order of priority is turning right, going straight, and turning left;

The left-hand rule: When there are multiple choices of heading, the order of priority is turning left, going straight, and turning right;

The central rule: When there are multiple choices of heading, the priority is turning towards the end.

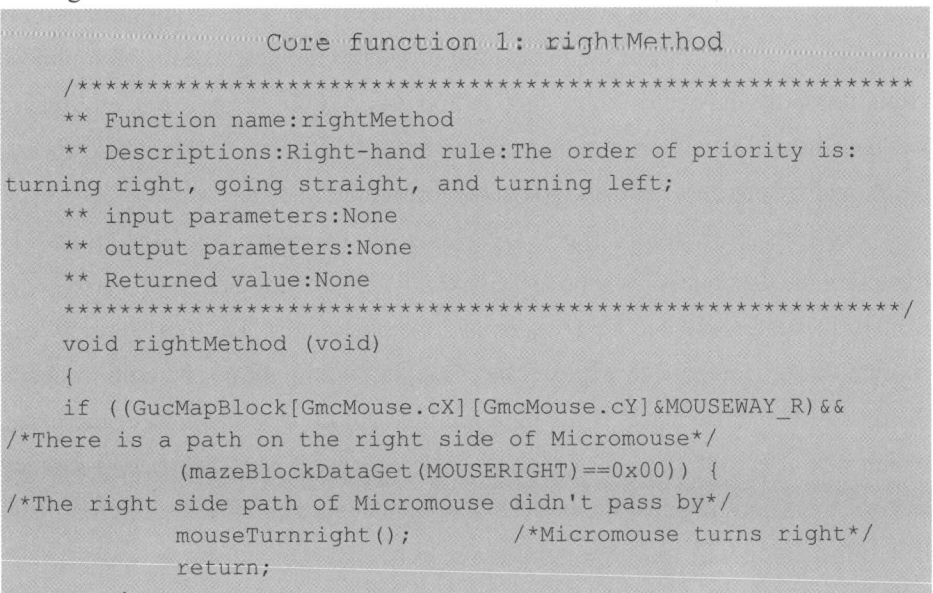

```
Core function 1: rightMethod
/****************************************************************
** Function name:rightMethod
** Descriptions:Right-hand rule:The order of priority is:
turning right, going straight, and turning left;
** input parameters:None
** output parameters:None
** Returned value:None
****************************************************************/
void rightMethod (void)
{
    if ((GucMapBlock[GmcMouse.cX][GmcMouse.cY]&MOUSEWAY_R)&&
/*There is a path on the right side of Micromouse*/
        (mazeBlockDataGet(MOUSERIGHT)==0x00)) {
/*The right side path of Micromouse didn't pass by*/
        mouseTurnright();        /*Micromouse turns right*/
        return;
    }
```

```
        if ((GucMapBlock[GmcMouse.cX][GmcMouse.cY] & MOUSEWAY_F) &&
/*There is a path on the front of Micromouse*/
            (mazeBlockDataGet(MOUSEFRONT)==0x00)) {
/*The front of Micromouse didn't pass by*/
            return;              /*Micromouse doesn't need to turn*/
        }
        if ((GucMapBlock[GmcMouse.cX][GmcMouse.cY]&MOUSEWAY_L) &&
/*There is a path on the left side of Micromouse*/
            (mazeBlockDataGet(MOUSELEFT )==0x00)) {
/*The left side path of Micromouse didn't pass by*/
            mouseTurnleft();              /*Micromouse turns left*/
            return;
        }
    }

                  Core function 2: leftMethod
    /****************************************************************
    ** Function name:leftMethod
    ** Descriptions:Left-hand rule:The order of priority is:
turning left, going straight, and turning right;
    ** input parameters:None
    ** output parameters:None
    ** Returned value:None
    ****************************************************************/
    void leftMethod (void)
    {
        if ((GucMapBlock[GmcMouse.cX][GmcMouse.cY]&MOUSEWAY_L) &&
/*There is a path on the left side of Micromouse*/
            (mazeBlockDataGet(MOUSELEFT )==0x00)) {
/*The left side path of Micromouse didn't pass by*/
            mouseTurnleft();              /*Micromouse turns left*/
            return;
        }
        if ((GucMapBlock[GmcMouse.cX][GmcMouse.cY] & MOUSEWAY_F)&&
/*There is a path on the front of Micromouse*/
            (mazeBlockDataGet(MOUSEFRONT)==0x00)) {
/*The front of Micromouse didn't pass by*/
            return;              /*Micromouse doesn't need to turn*/
        }
        if ((GucMapBlock[GmcMouse.cX][GmcMouse.cY] & MOUSEWAY_R)&&
/*There is a path on the right side of Micromouse*/
            (mazeBlockDataGet(MOUSERIGHT)==0x00)) {
/*The right side path of Micromouse didn't pass by*/
            mouseTurnright();              /*Micromouse turns right*/
            return;
        }
    }
```

```
                    Core function 3: centralMethod
    /*****************************************************************
    ** Function name:centralMethod
    ** Descriptions:Central rule:The central rule determines which
search rule to use based on the current position of Micromouse in
the maze.
    ** input parameters:None
    ** output parameters:None
    ** Returned value:None
    *****************************************************************/
    void centralMethod (void)
    {
        if (GmcMouse.cX&0x08) {
            if (GmcMouse.cY&0x08) {
                /*
                 *Micromouse is in the upper right corner of the maze.
                 */
                switch (GucMouseDir) {
                case UP:    /*Currently, Micromouse is facing up*/
                    leftMethod();           /*left-handrule*/
                    break;
                case RIGHT: /*Currently, Micromouse is facing right*/
                    rightMethod();          /*right-hand rule*/
                    break;
                case DOWN:   /*Currently, Micromouse is facing down*/
                    frontRightMethod();     /*Center right rule*/
                    break;
                case LEFT:   /*Currently, Micromouse is facing left*/
                    frontLeftMethod();      /*Center left rule*/
                    break;
                default:
                    break;
                }
            } else {
                /*
                 *Micromouse is in the lower right corner of the maze.
                 */
                switch (GucMouseDir) {
                case UP:       /*Currently, Micromouse is facing up*/
                    frontLeftMethod();      /*Center left rule*/
                    break;
                case RIGHT: /*Currently, Micromouse is facing right*/
                    leftMethod();           /*left-hand rule*/
                    break;
                case DOWN:   /*Currently, Micromouse is facing down*/
                    rightMethod();          /*right-hand rule*/
```

```c
                break;
        case LEFT:    /*Currently, Micromouse is facing left*/
            frontRightMethod();      /*Center right rule*/
            break;
        default:
            break;
        }
    }
} else {
    if (GmcMouse.cY&0x08) {
        /*
         * Micromouse is in the upper left corner of the maze.
         */
        switch (GucMouseDir) {
        case UP:         /*Currently, Micromouse is facing up*/
            rightMethod();             /*right-hand rule*/
            break;
        case RIGHT:    /*Currently, Micromouse is facing right*/
            frontRightMethod();      /*Center right rule*/
            break;
        case DOWN:    /*Currently, Micromouse is facing down*/
            frontLeftMethod();       /*Center left rule*/
            break;
        case LEFT:    /*Currently, Micromouse is facing left*/
            leftMethod();              /*left-hand rule*/
            break;
        default:
            break;
        }
    } else {
        /*
         * Micromouse is in the lower left corner of the maze.
         */
        switch (GucMouseDir) {
        case UP:         /*Currently, Micromouse is facing up */
            frontRightMethod();      /*Center right rule*/
            break;
        case RIGHT:    /*Currently, Micromouse is facing right*/
            frontLeftMethod();       /*Center left rule*/
            break;
        case DOWN:    /*Currently, Micromouse is facing down*/
            leftMethod();              /*left-hand rule*/
            break;
        case LEFT:    /*Currently, Micromouse is facing left*/
            rightMethod();             /*right-hand rule*/
            break;
        default:
```

```
                        break;
                    }
                }
            }
        }
```

Reflection and Summary

(1) What are the advantages and disadvantages of the left-hand rule, the right-hand rule and the central rule?

(2) The search rules of Micromouse are all composed of the left-hand rule and the righ-hand rule. According to certain rules, the maze is divided into several parts. Different turns are selected when Micromouse has different positions and orientations.

Chapter 3 Advanced Skills and Competitions | 065

Project 2

The Principle of Path Planning

Learning objectives

(1) Learning about the information storage method of a maze.

(2) Learning about the principle of step map and turning weighted, and try to draw the optimal path on a competition maze.

Task 1 The Information Storage Method of a Maze

For path planning, it is necessary to record the walls' information of all cells at first. Obviously, it is an effective method to establishing a two-dimensional array to define the coordinates of the entire maze. Each cell is defined as a coordinate, and the corresponding walls' information is stored in the established two-dimensional array.

When Micromouse reaches a cell, it should record the data of the wall according to the sensors detection results. In order to facilitate management and save storage space, the lower four bits of each byte variable are used to store the wall data around a cell. The maze has 16×16 cells in total, so a 16×16 two-dimensional array variables can be defined to save the whole maze wall's information, as shown in Fig. 3–2–1.

First, initialize the data of all the maze walls to 0. At least one side of a cell that Micromouse has walked through has no wall， so that the data are not all 0, therefore you can determine whether the cell has been searched by whether the data stored in the cell are 0. Information storage method is shown as Table 3–2–1.

066 | Micromouse Design Principles and Production Process (Intermediate)

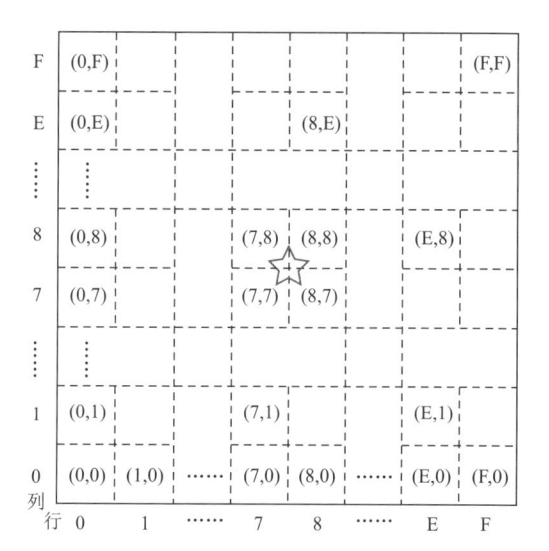

Fig. 3-2-1   Maze coordinate definition

Table 3-2-1   Information storage method

Variable	Direction	One wall or not
bit0	Up 0	1: No, 0: Yes
bit1	Right 1	1: No, 0: Yes
bit2	Down 2	1: No, 0: Yes
bit3	Left 3	1: No, 0: Yes
bit7- bit4		Reserved

Task 2   The Step Map Making Method

Assuming that Micromouse has searched the whole maze or only part of the maze including the start and the destination, and has recorded the walls data of each cell it has passed, then how can it find the optimal path from the start to the destination? The following introduces the concept and the method of step map making.

Step map is widely used in the field of geography and meteorology. It can mark the area range of the same altitude or the range and size of air pressure.

Then the step map can be used on the maze map to calculate the distance between each maze cell and the destination, until get the distance between the start and the destination. By sorting the distance of each cell to the destination from large to small, we can find the optimal path in the maze.

● Video

The method of finding the optimal path—step map

Chapter 3　Advanced Skills and Competitions | 067

Marking the starting as 1. According to the walls' information of each cell, the shortest steps to the start are marked on the cell so that the optimal path from any coordinate to the start is obtained, as shown in Fig. 3–2–2.

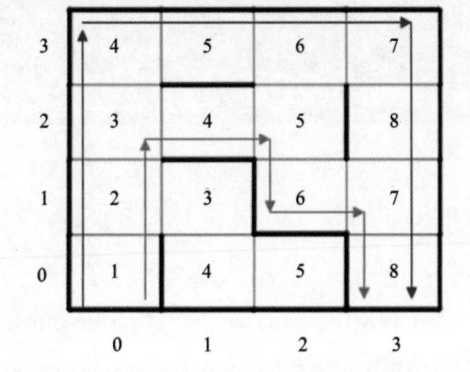

Fig. 3–2–2　The sketch map of step map

Reflection and Summary

(1) How does Micromouse record the walls' information of each cell?

(2) How does Micromouse plan the optimal path?

(3) Micromouse needs to slow down and accelerate when it turns. Forward going have priority over turns, and the long straights have priority over short straights.

068 | Micromouse Design Principles and Production Process (Intermediate)

Project 3

Micromouse Program Design

Learning objectives

Learn the programming of Micromouse.

The flexibility and intelligence of TQD-Micromouse-JD are not only depend on the structure and performance of hardware, but also depend on the programming integrity. The more intelligent Micromouse is, the more complex the programming. In the Micromouse programming, the overall structure of the programming can be simply divided into two parts, namely the bottom driver program and the top algorithm program.

The bottom driver program is mainly for the realization of some basic functions of Micromouse, such as controlling its going-forward N cells in a straight line, measuring the distance it has moved forward, turning 90° to the right or left, preventing collision with the wall, and detecting the walls' information around the cell, etc.

The top algorithm program is mainly the intelligent algorithm of Micromouse, such as determining the action of Micromouse according to the maze information, remembering the map of the maze that it has walked, and finding the optimal path to the destination, etc.

Task 1 Attitude Program Control

After the operation of TQD-Micromouse-JD, it will process a lot of information and switch its states.

1) Waiting state

In this state, TQD-Micromouse-JD stops at the start and waits for the start command. At the same time, the sensors' detection results and battery voltage are displayed in real time, which are convenient for debugging the sensitivity of

Chapter 3 Advanced Skills and Competitions | 069

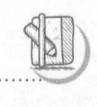

the sensors and replacing the battery. When the button that controls the start is pressed, TQD-Micromouse-JD will enter the starting state.

2) Starting state

In this state, TQD-Micromouse-JD will determine whether the start coordinate is (0,0) or (F,0) according to the direction of the first turning. Flow chart of the start coordinate judgment is shown as Fig. 3-3-1.

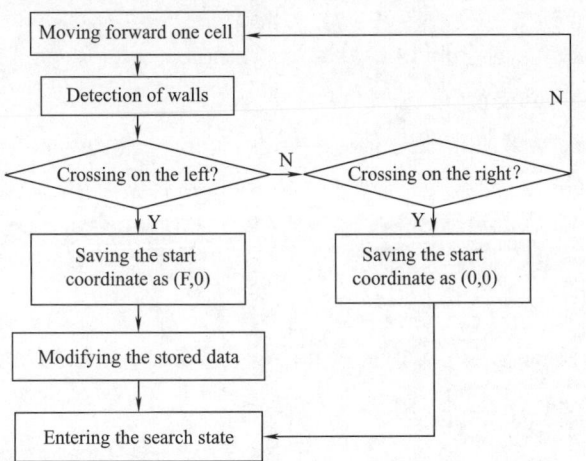

Fig. 3-3-1　Flow chart of the start coordinate judgment

3) Searching state

In this state, the task of TQD-Micromouse-JD is to explore and memorize the maze map. Here we use the right-hand rule to search the whole maze. Flow chart of the maze searching is shown as Fig. 3-3-2.

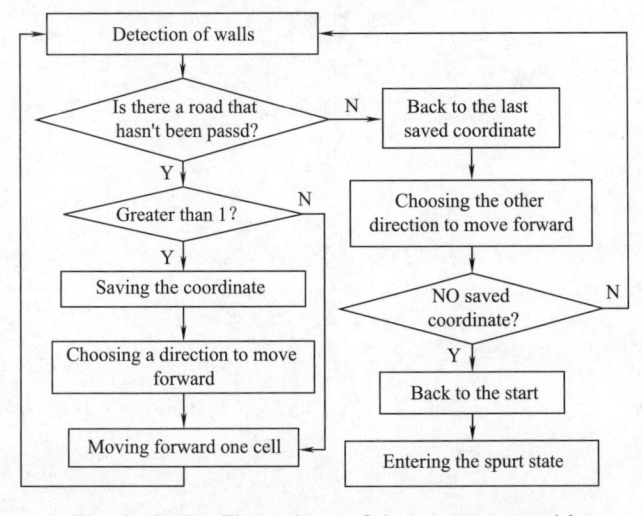

Fig. 3-3-2　Flow chart of the maze searching

4）Spurting state

After the maze search, Micromouse will find an optimal path to spurt to the destination according to the algorithm. It will return to the start after the spurt.

Task 2   Analysis of the Main Program Structures

Here are a few main functions that Micromouse calls when it runs.

1) Maze searching program

Micromouse will run the maze Search function at every coordinate, and detect the walls' information around and store them.

```
                    Core function 1: mazeSearch
/**********************************************************
** Function name:mazeSearch
** Descriptions:Moving forward N cells.
** input parameters:iNblock(Number of cells move forward)
** output parameters:None
** Returned value:None
**********************************************************/
void mazeSearch(void)
{
    char cL=0, cR=0, cCoor=1;
    if (__GmLeft.cState) {
        cCoor=0;
    }                              /*Setting up the running task*/
    __GucMouseState=__GOAHEAD;
    __GiMaxSpeed=SEARCHSPEED;
    __GmRight.cDir=__MOTORGOAHEAD;
    __GmLeft.cDir=__MOTORGOAHEAD;
    __GmRight.uiPulse=MAZETYPE*ONEBLOCK;
    __GmLeft.uiPulse=MAZETYPE*ONEBLOCK;
    __GmRight.cState=__MOTORRUN;
    __GmLeft.cState=__MOTORRUN;

    while (__GmLeft.cState!=__MOTORSTOP) {
        if (__GmLeft.uiPulseCtr>=ONEBLOCK) {
                        /*Judging whether complete one cell*/
            __GmLeft.uiPulse-=ONEBLOCK;
            __GmLeft.uiPulseCtr-=ONEBLOCK;
            if (cCoor) {
                __mouseCoorUpdate(); /*Updating coordinate*/
            } else {
                cCoor=1;
            }
```

Chapter 3　Advanced Skills and Competitions | 071

```c
            }
         if (__GmRight.uiPulseCtr >=ONEBLOCK) {
/*Judging whether complete one cell or not.*/
            __GmRight.uiPulse-=ONEBLOCK;
            __GmRight.uiPulseCtr-=ONEBLOCK;
         }
         if (__GucDistance[__FRONT]) {
/*If there is a wall on the front, jump out of the program*/
            __GmRight.uiPulse=__GmRight.uiPulseCtr+70;
            __GmLeft.uiPulse=__GmLeft.uiPulseCtr +70;
            while (1) {
               if ((__GmLeft.uiPulseCtr+20)>__GmLeft.uiPulse) {
                  goto End;
               }
            }
         }
         if (cL) {   /*Whether allowing detection on the left or not*/
            if ((__GucDistance[ __LEFT]&0x01)==0) {
      __GmRight.uiPulse=__GmRight.uiPulseCtr+74;
/*If there is a branch on the left, jump out of the program*/
               __GmLeft.uiPulse=__GmLeft.uiPulseCtr +74;
               while ((__GucDistance[ __LEFT] & 0x01)==0) {
                  if ((__GmLeft.uiPulseCtr+20)>__GmLeft.uiPulse) {
                     goto End;
                  }
               }
               __GmRight.uiPulse=MAZETYPE *ONEBLOCK;
               __GmLeft.uiPulse=MAZETYPE *ONEBLOCK;
            }
         } else {
   if (__GucDistance[ __LEFT]&0x01) {          '
/*When there is a wall on the left, it is allowed to detect the left side*/
               cL=1;
            }
         }
         if (cR) {   /*Whether allowing detection on the right or not*/
      if ((__GucDistance[__RIGHT]&0x01)==0) {
/*If there is a branch on the right, jump out of the program*/
               __GmRight.uiPulse=__GmRight.uiPulseCtr+74;
               __GmLeft.uiPulse=__GmLeft.uiPulseCtr +74;
               while ((__GucDistance[ __RIGHT] & 0x01)==0) {
                  if ((__GmLeft.uiPulseCtr+20)>__GmLeft.uiPulse) {
                     goto End;
                  }
               }
               __GmRight.uiPulse=MAZETYPE *ONEBLOCK;
               __GmLeft.uiPulse=MAZETYPE *ONEBLOCK;
```

```
                    }
            } else {
                if (__GucDistance[__RIGHT]&0x01) {
/*When there is a wall on the right, it is allowed to defect the right side*/
                    cR=1;
                }
            }
        }
    End:__mouseCoorUpdate();                    /*Updating coordinate*/
    }
```

2) Step map making program

Combining its own rules, Micromouse integrates the walls' information of all coordinates, and plans the optimal path.

```
                    Core function2: mapStepEdit
    /****************************************************************
    ** Function name:mapStepEdit
    ** Descriptions:
    ** input parameters:uiX(The X-coordinate of the destination)
    **                  uiY(The Y-coordinate of the destination)
    ** output parameters:GucMapStep[][](Step values of each coordinate)
    ** Returned value:None
    ****************************************************************/
    void mapStepEdit (char  cX, char  cY)
    {
        uchar n=0;          /*Counting the number of crossings*/
        uchar ucStep=1;     /*Step map value*/
        uchar ucStat=0;     /*Counting the number of directions can
    move forward.*/
        uchar i,j;

        GmcStack[n].cX=cX;   /*Saving start X-coordinate in stack*/
        GmcStack[n].cY=cY;   /*Saving start Y-coordinate in stack*/
        n++;
        /*
         *  Step map value initialization
         */
        for (i=0; i<MAZETYPE; i++) {
            for (j=0; j<MAZETYPE; j++) {
                GucMapStep[i][j]=0xff;
            }
        }
        /*
         *  Making step map, until all the data in stack are processed
```

```c
        */
    while (n) {
        GucMapStep[cX][cY]=ucStep++;   /*Filling in step value*/
/*Counting the direction that the current coordinate can move forward*/
        ucStat=0;
        if ((GucMapBlock[cX][cY]&0x01)&&  /*There is a way upper*/
        (GucMapStep[cX][cY+1] > (ucStep))) {
/*The upper step value is greater than the plan setting value*/
            ucStat++;    /*Forward direction numbers plus 1*/
        }
        if ((GucMapBlock[cX][cY]&0x02) &&
/*There is a way on the right*/
            (GucMapStep[cX+1][cY]>(ucStep))) {
/*The right step value is greater than the plan setting value*/
            ucStat++;    /*Forward direction numbers plus 1*/
        }
        if ((GucMapBlock[cX][cY]&0x04) &&  /*There is a way lower*/
        (GucMapStep[cX][cY-1]>(ucStep))) {
/*The lower step value is greater than the plan setting value*/
            ucStat++;    /*Forward direction numbers plus 1*/
        }
        if ((GucMapBlock[cX][cY]&0x08) &&
/*There is a way on the left*/
        (GucMapStep[cX-1][cY]>(ucStep))) {
/*The left step value is greater than the plan setting value*/
            ucStat++;    /*Forward direction numbers plus 1*/
        }
        /*If there is no direction to go forward, jumping
to the nearest saved branch point. Otherwise, selecting any one
direction to go forward.*/
        if (ucStat==0) {
            n--;
            cX=GmcStack[n].cX;
            cY=GmcStack[n].cY;
            ucStep=GucMapStep[cX][cY];
        } else {
            if (ucStat > 1) {
/*If there are multiple directions can move forward, save the coordinate.*/
                GmcStack[n].cX=cX;  /*Saving X coordinate in stack*/
                GmcStack[n].cY=cY;  /*Saving Y coordinate in stack*/
                n++;
            }
            /*
             * Choosing any one direction to move forward.
             */
            if ((GucMapBlock[cX][cY]&0x01) &&
/*There is a way upper*/
```

```c
                (GucMapStep[cX][cY+1]>(ucStep))) {
/*The upper step value is greater than the plan setting value*/
                    cY++;                    /*Modifying coordinate*/
                    continue;
                }
                if ((GucMapBlock[cX][cY]&0x02) &&
/*There is a way on the right*/
            (GucMapStep[cX+1][cY]>(ucStep))) {
/*The right step value is greater than the plan setting value*/
                    cX++;                    /*Modifying coordinate*/
                    continue;
                }
                if ((GucMapBlock[cX][cY]&0x04) &&
/*There is a way lower*/
            (GucMapStep[cX][cY-1]>(ucStep))) {
/*The lower step value is greater than the plan setting value*/
                    cY--;                    /*Modifying coordinate*/
                    continue;
                }
            if ((GucMapBlock[cX][cY]&0x08) &&
/*There is a way on the left*/
            (GucMapStep[cX-1][cY]>(ucStep))) {
/*The left step value is greater than the plan setting value*/
                    cX--;                    /*Modifying coordinate*/
                    continue;
                }
            }
        }
    }
    if (cDirTemp==GucMouseDir) {
/*Giving priority to the direction that do not need turning*/
                    cNBlock++;               /*Going ahead one cell*/
                    cY++;
                    continue;                /*Skipping this loop*/
                }
            }
            if ((GucMapBlock[cX][cY]&0x02) &&
/*There is a way on the right*/
                (GucMapStep[cX+1][cY]<ucStep)) {
/*The right step value is a little smaller*/
                cDirTemp=RIGHT;             /*Recording the direction*/
                if (cDirTemp==GucMouseDir) {
/*Giving priority to the direction that do not need turning*/
                    cNBlock++;               /*Going ahead one cell*/
                    cX++;
                    continue;                /*Skipping this loop*/
                }
```

Chapter 3 Advanced Skills and Competitions | 075

```c
            }
            if ((GucMapBlock[cX][cY]&0x04) &&
/*There is a way lower*/
                (GucMapStep[cX][cY-1]<ucStep)) {
/*The lower step value is a little smaller*/
                cDirTemp=DOWN;          /*Recording the direction*/
                if (cDirTemp==GucMouseDir) {
/*Giving priority to the direction that do not need turning*/
                    cNBlock++;          /*Going ahead one cell*/
                    cY--;
                    continue;           /*Skipping this loop*/
                }
            }
            if ((GucMapBlock[cX][cY]&0x08) &&
/*There is a way on the left*/
                (GucMapStep[cX-1][cY]<ucStep)) {
/*The left step value is a little smaller*/
                cDirTemp=LEFT;          /*Recording the direction*/
                if (cDirTemp==GucMouseDir) {
/*Giving priority to the direction that do not need turning*/
                    cNBlock++;          /*Going ahead one cell*/
                    cX--;
                    continue;           /*Skipping this loop*/
                }
            }
            cDirTemp=(cDirTemp+4-GucMouseDir)%4;
/*Calculating direction offset*/
            if (cNBlock) {
                mouseGoahead(cNBlock); /*Going ahead cNBlock cells*/
            }
            cNBlock=0;                      /*Task reset*/
            switch (cDirTemp) {    /*Controling Micromouse to turn.*/
```

3) Jumping to specified coordinate program

The purpose of this program block is to control Micromouse to move forward to the designated coordinate in the shortest path. Of course, the premise of this function is that Micromouse has searched this coordinate.

```c
                Core function 3: objectGoTo
    /***************************************************************
    ** Function name:objectGoTo
    ** Descriptions:Moving to the specified coordinate
    ** input parameters:cXdst(The X-coordinate of the destination)
    **                  cYdst(The Y-coordinate of the destination)
```

```
** output parameters:None
** Returned value:None
*************************************************************/
void objectGoTo (char  cXdst, char  cYdst)
{
    uchar ucStep=1;
    char  cNBlock=0, cDirTemp;
    char cX,cY;
    cX=GmcMouse.cX;
    cY=GmcMouse.cY;
    mapStepEdit(cXdst,cYdst);  /*Making step map*/
    /*
     *  According to the step map value, moving to the destination
     */
    while ((cX!=cXdst)||(cY!=cYdst)) {
        ucStep=GucMapStep[cX][cY];
        /*
         *  Choosing one direction that the step map value is
smaller than the current one to move forward
         */
        if ((GucMapBlock[cX][cY]&0x01)&&  /*There is a way upper*/
            (GucMapStep[cX][cY+1]<ucStep)) {
/*The upper step value in a little smaller*/
            cDirTemp=UP;           /*Recording the direction*/
    if (cDirTemp==GucMouseDir) {
/*Giving priority to the direction that do not need turning*/
            cNBlock++;          /*Going ahead one cell*/
            cY++;
            continue;           /*Skipping this loop*/
            }
        }
        if ((GucMapBlock[cX][cY]&0x02) &&
/*There is a way on the right*/
            (GucMapStep[cX+1][cY]<ucStep)) {
/*The right step value is a little smaller*/
            cDirTemp=RIGHT;        /*Recording the direction*/
            if (cDirTemp==GucMouseDir) {
/*Giving priority to the direction that do not need turning*/
            cNBlock++;          /*Going ahead one cell*/
            cX++;
            continue;           /*Skipping this loop*/
            }
        }
        if ((GucMapBlock[cX][cY]&0x04)&&  /*There is a way lower*/
            (GucMapStep[cX][cY-1]<ucStep)) {
/*The lower step value is a little smaller*/
            cDirTemp=DOWN;          /*Recording the direction*/
```

```c
                if (cDirTemp==GucMouseDir) {
/*Giving priority to the direction that do not need turning*/
                cNBlock++;          /*Going ahead one cell*/
                cY--;
                continue;           /*Skipping this loop*/
            }
        }
        if ((GucMapBlock[cX][cY]&0x08)&&
/*There is a way on the left*/
            (GucMapStep[cX-1][cY]<ucStep)) {
/*The left step value is a little smaller*/
            cDirTemp=LEFT;          /*Recording the direction*/
            if (cDirTemp==GucMouseDir) {
/*Giving priority to the direction that do not need turning*/
                cNBlock++;          /*Going ahead one cell*/
                cX--;
                continue;           /*Skipping this loop*/
            }
        }
        cDirTemp=(cDirTemp+4-GucMouseDir)%4;
/*Calculating direction offset*/
        if (cNBlock) {
            mouseGoahead(cNBlock); /*Going ahead cNBlock cells*/
        }
        cNBlock=0;                      /*Task reset*/
        switch (cDirTemp) {     /*Controling Micromouse to turn.*
        case 1:
            mouseTurnright();
            break;
        case 2:
            mouseTurnback();
            break;
        case 3:
            mouseTurnleft();
            break;

        default:
            break;
        }
    }
    /*
     *  Judging whether the task is completed
     */
    if (cNBlock) {
        mouseGoahead(cNBlock);
    }
}
```

4) Counting the number of branches not searched program

This program is used to count the total number of unsearched branches around the specified coordinate so that the system can choose searching strategy. The program is as below:

```
                  Core function 4: crosswayCheck
/*********************************************************************
** Function name:crosswayCheck
** Descriptions:Counting the number of directions that have
not passed by
** input parameters:ucX(The X-coordinate of the current point)
**                  ucY(The Y-coordinate of the current point)
** output parameters:None
** Returned value:ucCt(The directions that have not passedby)
*********************************************************************/
uchar crosswayCheck (char  cX, char  cY)
{
    uchar ucCt=0;
   if ((GucMapBlock[cX][cY]&0x01) &&
/*Absolute direction, there's a way up here in the maze*/
    (GucMapBlock[cX][cY+1])==0x00) {
/*Absolute direction, the upper path in the maze didn't pass by*/
        ucCt++;            /*Forward direction numbers plus 1*/
    }
      if ((GucMapBlock[cX][cY]&0x02) &&
/*Absolute direction, there's a way on the right in the maze*/
       (GucMapBlock[cX+1][cY])==0x00) { *Absolute direction,
the right path in the maze didn't pass by*/
            ucCt++;            /*Forward direction numbers plus 1*/
    }
      if ((GucMapBlock[cX][cY]&0x04) &&
/*Absolute direction, there's a way down here in the maze*/
       (GucMapBlock[cX][cY-1])==0x00) {
/*Absolute direction, the lower path in the maze didn't pass by*/
       ucCt++;            /*Forward direction numbers plus 1*/
    }
      if ((GucMapBlock[cX][cY]&0x08) &&
/*Absolute direction, there's a way on the left in the maze*/
       (GucMapBlock[cX-1][cY])==0x00) {
/*Absolute direction, the left path in the maze didn't pass by*/
        ucCt++;            /*Forward direction numbers plus 1*/
    }
    return ucCt;
}
```

Chapter 3 Advanced Skills and Competitions | 079

5) TQD-Micomouse-JD main program

```c
/****************************************************************
** Function name:main
** Descriptions:main function
** input parameters:None
** output parameters:None
** Returned value:None
****************************************************************/
main (void)
{
    uchar n=0; /*The number of coordinates with multiple branches*/
    uchar ucRoadStat=0;
        /*Counting the number of directions can move forward*/
    uchar ucTemp=0;/*Used for coordinate conversion in START state*/
    mouseInit();        /*Underlying driver initialization*/
    Init_7289();        /*Initialization of the display module*/
    while (1) {
        switch (GucMouseTask) {
        case WAIT:
            sensorDebug();
            voltageDetect();
            delay(100000);
            if (keyCheck()==true) {
                Reset_7289();
                GucMouseTask=START;
            }
            break;
    case START:                     /*Judging the x coordinate of
the start point of Micromouse*/
        mazeSearch();                 /*Searching forward.*/
    if (GucMapBlock[GmcMouse.cX][GmcMouse.cY]&0x08) {
/*Judging whether there is an exit on the left of Micromouse.*/
                if (MAZETYPE==8) {
/*Modifying the destination coordinates of the quarter maze.*/
                    GucXGoal0=1;
                    GucXGoal1=0;
                }
        GucXStart=MAZETYPE-1;
/*Modifying the x coordinate of the start point of Micromouse*/
        GmcMouse.cX=MAZETYPE-1;
/*Modifying the x coordinate of the start point of Micromouse*/
                /*
                 *  Because the default start coordinate is
(0,0), now we need to convert the recorded wall data.
                 */
                ucTemp=GmcMouse.cY;
                do {
```

```c
                        GucMapBlock[MAZETYPE-1]
[ucTemp]=GucMapBlock[0][ucTemp];
                        GucMapBlock[0][ucTemp]=0;
                    }while (ucTemp--);
/*Saving the start coordinates in OFFSHOOT[0]*/
                    GmcCrossway[n].cX=MAZETYPE-1;
                    GmcCrossway[n].cY=0;
                    n++;
                    GucMouseTask=MAZESEARCH;            /*Changing
state to search state*/
                }
                if (GucMapBlock[GmcMouse.cX][GmcMouse.cY]&0x02) {
/*Judging whether there is an exit on the right of Micromouse.*/
                    GmcCrossway[n].cX=0;
                    GmcCrossway[n].cY=0;
                    n++;
                    GucMouseTask=MAZESEARCH;    /*Changing state
to search state*/
                }
            break;
        case MAZESEARCH:
            if (((GmcMouse.cX==GucXGoal0)&&(GmcMouse.
cY==GucYGoal0))||((GmcMouse.cX==GucXGoal0)&&(GmcMouse.cY==GucYGoal1))
    ||((GmcMouse.cX==GucXGoal1)&&(GmcMouse.
cY==GucYGoal0))||((GmcMouse.cX==GucXGoal1)&&(GmcMouse.cY==GucYGoal1)))
            {
                mouseTurnback();
                objectGoTo(GucXStart,GucYStart);
                mouseTurnback();
                GucMouseTask=SPURT;
                break;
            }
            else{
        ucRoadStat=crosswayCheck(GmcMouse.cX,GmcMouse.cY);
/*Counting the number of branches that can go ahead. */
                if (ucRoadStat)  /*There is at least one path can
go ahead*/
                {
                    if (ucRoadStat>1) {
/*There are multiple forward directions, save the coordinates*/
                        GmcCrossway[n].cX=GmcMouse.cX;
                        GmcCrossway[n].cY=GmcMouse.cY;
                        n++;
                    }
                crosswayChoice();
/*Using the right-hand rule to search and select the direction to go ahead*/
                    mazeSearch();     /*Going ahead one cell.*/
```

```
                }
                  else if(ucRoadStat==1)
                  {
                  crosswayChoice();
/*Using the right-hand rule to search and select the direction to go ahead*/
                      mazeSearch();
                  }
                  else
      {    /*There's no way forward. Go back to the nearest branch*/
                      mouseTurnback();
                      n=n-1;
                      objectGoTo(GmcCrossway[n].cX,GmcCrossway[n].cY);

              ucRoadStat = crosswayCheck(GmcMouse.cX,GmcMouse.cY);
                      if (ucRoadStat>1) {
                          GmcCrossway[n].cX=GmcMouse.cX;
                          GmcCrossway[n].cY=GmcMouse.cY;
                          n++;
                      }
                      crosswayChoice();
                      mazeSearch();
                  }
                }
                break;
            case SPURT:
            mouseSpurt();
/*Spurting to the destination with the optimal path*/
            objectGoTo(GucXStart,GucYStart); /*Back to the start*/
                mouseTurnback();
/*Turning back and return to the starting attitude*/
                while (1) {
                    if (keyCheck()==true) {
                        break;
                    }
                    sensorDebug();
                    delay(20000);
                }
                break;
            default:
                break;
            }
        }
    }
```

Reflection and Summary

(1) What are the relationships among the various stages of Micromouse programming?

(2) What are the main functions of Micromouse used for?

(3) The Micromouse program is executed by several main functions working together. Only when all the functions are debugged accurately, Micromouse can successfully reach the destination through the maze.

Appendix

Micromouse Design Principles and Production Process (Intermediate)

Appendix A

Micromouse Competition Going Popular in the World

Since 2019, it has been the most prosperous and fruitful period in the history of Micromouse competition.The International Micromouse Competition is held around the world, as shown in Fig.A–1.

In January, the International Micromouse Competition was held in Bombay, India.

In March, the APEC International Micromouse Competition was held in California, US.

In April, the International Micromouse Competition was held in Gondomar, Portugal.

In May, the IEEE Micromouse International Invitational Competition was held in Tianjin, China.

In June, the International Micromouse Competition was held in London, UK.

In August, the International Micromouse Competition was held in Chile.

In October, the International Micromouse Competition was held in Egypt.

In November, All Japan Micromouse International Competition was held in Tokyo, Japan.

Micromouse Competition Going Popular in the World

The International Micromouse Competition will become a booster of global higher education, vocational education, general education and technological innovation and integrated development of production and education. With the rapid development of artificial intelligence Micromouse competition, the education field timely introduced international well-known competitions to improve students' professional comprehensive ability, master the experience of practice and innovation, and help the integrated development of industry and

education, and cultivate more excellent seed talents for the industry, profession and enterprise.

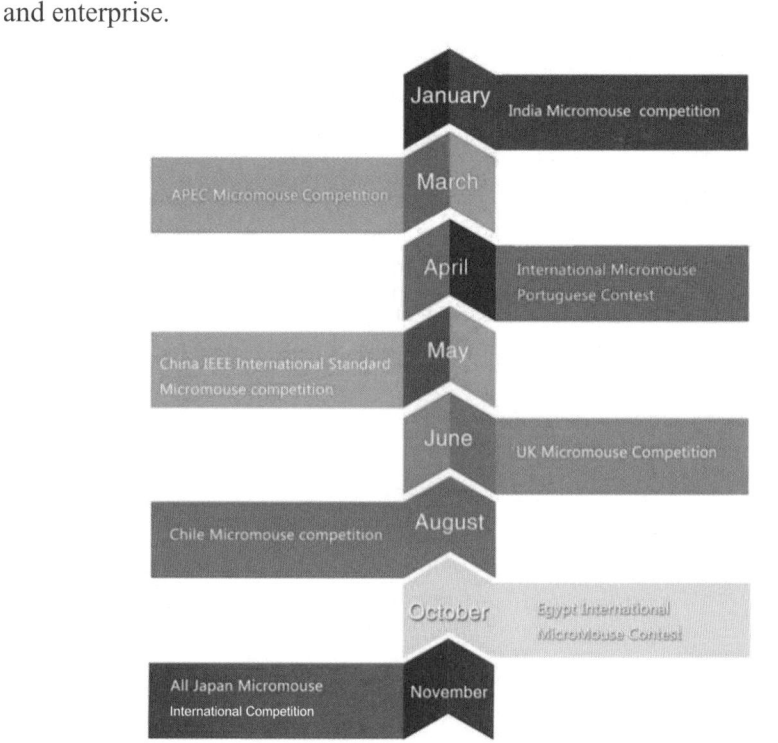

Fig. A–1   International Micromouse Competitions

1) China IEEE Micromouse International Invitational Competition

In 2009, Tianjin Qicheng Science and Technology Co., Ltd. introduced Micromouse competition into China, and carried out localized innovation and reform in the IEEE International Standard Micromouse Competition, which played a leading role in satisfying industrial optimization and upgrading, broadening international vision, gaining practice and innovation experience, and cultivating high-tech talents.

From 2016 to 2019, IEEE Micromouse International Invitational Competition has been successfully held for four times. The competition is hosted by Tianjin Municipal Education Commission and organized by Tianjin Qicheng Science and Technology Co. Ltd. and Tianjin Bohai Vocational Technical College, as shown in Fig. A–2.

Fig. A–2 The IEEE Micromouse International Invitational Competition in China has been held since 2016

At present, IEEE Micromouse International Invitational Competition in China has set up "middle school, vocational college, bachelor's degree, master's degree and occupation" five competition group. It aims to improve the social participation and professional coverage of the competition. Micromouse has become an important carrier of systematic training and education. It fully embodies the combination of optical and electrical, software and hardware, control and machinery. While deducing the concept of "engineering" course, it extends and expands the concept of "innovative" course, which makes the content of students' learning and the teaching method of teachers have a new connotation, and truly focuses on the cultivation of comprehensive quality to create happy quality education.

IEEE Micromouse International Invitational Competition in China has the following characteristics.

(1) Participants: Facing not only college students, but also primary school, middle school and vocational workers, reflecting the characteristics of through training and lifelong education. It also includes international Micromouse professional players and previous international Micromouse competition winners.

(2) Maze site: There are 8×8 Micromouse maze sites for primary and secondary schools, and also have 16×16 full size classical Micromouse maze site for colleges and universities. Other than that, there is also a 25×32 half size Micromouse maze for elite players. Reflecting the extensibility of the competition, it takes the Micromouse competition as the core form, and students from different learning stages can participate in the competition.

(3) Competition events: There are not only Micromouse competition,

Video

China IEEE Micromouse International Invitational Competition

but also Robotracer race, which reflects the competition is both technical and engineering, and show the idea of engineering application oriented competition.

(4) Competition rules: A comparison of the similarities and differences in competition rules of general education, vocational education, higher education, vocational elite, as shown in Table A–1.

Table A–1   A comparison of similarities and differences in competition rules

Entry category	General education	Vocational education	Higher education	Vocational elite
Competition form	(1) Online debugging of APP. (2) Graphical programming. (3) Application of IOT intelligent sensing technology. (4) 8×8 maze race	(1) The oretical knowledge assessment. (2) According to the referee the task programming and implement the corresponding function (3) On-site technical defense. (4) 16×16 classical maze racing	(1) DIY appearance and structure mechanical design. (2) Hardware technology innovation. (3) Program algorithm innovation. (4) 16 × 16 classical maze racing	(1) DIY appearance and structure mechanical design. (2) Hardware technology innovation. (3) Program algorithm innovation. (4) 25×32 half size maze racing
Competition content	(1) Assembly task 10%. (2) Debugging task 40%. (3) Racing task 50%	(1) Theoretical assessment 20%. (2) Innovation 30%. (3) Speed race 50%	(1) Innovation 20%. (2) Speed race 80%	Speed race 100%

International Micromouse experts on-site training guidance is shown in Fig. A–3.

Fig. A–3   International Micromouse experts on-site training guidance

2) US APEC International Micromouse Competition

In 1977, the first exciting Micromouse competition was held in New York, US. It was co-sponsored by IEEE and APEC. Thus, the most influential international American APEC world Micromouse Competition was born. Known as one of the world's three major Micromouse competitions, it has held 34 competitions until 2019.

The official website of APEC: http://www.apec-conf.org/.

American Micromouse enthusiasts website: http://Micromouse usa.com/, as shown in Fig.A–4.

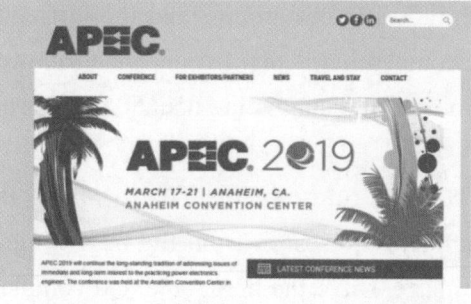 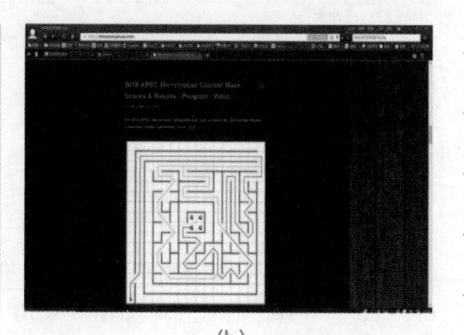

(a)   (b)

Fig. A–4   Screenshots of official website of American APEC International Micromouse Competition

Competition time: Between February and April every year.

Competition venue: Varies annually (previous venues include North Carolina, Texas, Florida, California, etc.). Every year, different countries such as the United States, the United Kingdom, Japan, South Korea, Singapore, India, China have actively participated in the competition, as shown in Fig. A–5.

Fig. A–5   The 30th American APEC International Micromouse Competition

3) UK International Micromouse Competition

Since 1980, UK International Micromouse Competition has become one of the internationally well-known Micromouse competitions.

Competition time: June every year.

Competition venue: Birmingham City University.

The competition is sponsored by UK Micromouse and Robotics Society. The

characteristic of the Micromouse competition in UK is that it does not restrict anyone to participate,whether you are from middle school, university or social personnel. All the players are divided into different groups, and the difficulty of maze is adjusted appropriately. The competition is divided into line follower, wall follower, maze solver and other projects, attracting more than 50 teams from more than 10 countries of the world.

UK Micromouse Competition scoring rules: In the 16×16 maze, the participating Micromouse need to complete the search from the beginning to the end and the traversal of the whole maze, solve the best route and complete the sprint from the beginning to the end. Scoring time = search time (time used to find the end for the first time) / 30 + sprint time (high-speed sprint with the shortest path from the starting point to the end) + penalty time (3s/time for knocking into the wall).

The offical website: https://ukmars.org/index.php/Main_Page, as shown in Fig. A–6.

Fig. A–6　Screenshot of official website of UK Micromouse Competition

4) All Japan Micromouse International Competition

All Japan Micromouse International Competition has been held 40 times from 1980 to 2019.

Competition time: November or December every year.

Competition venue: Tokyo, Japan.

The official Website: http://www.ntf.or.jp/mouse/Micromouse 2018/index. html, as shown in Fig. A–7.

Every year, Micromouse teams from more than 20 countries such as the United States, Britain, Japan, Singapore, China, Mongolia, Chile, Portugal participate in the competition, as shown in Fig. A–8.

Video

All Japan
Micromouse
International
Competition

Fig. A–7　Screenshot of official website of All Japan Micromouse International Competition

Fig. A–8　Prize-giving of the 39th All Japan Micromouse International Competition

The competition consists of classic Micromouse event, half size Micromouse event and Robotrace event. The team consists of middle school students, college students and vocational elites. According to statistics, there are more than 300 teams. All Japan Micromouse International Competition can be said to represent today's international Micromouse technology field with the highest level and the strongest technology, so it has attracted much attention.

5) Chile Micromouse International Competition

The Ministry of Foreign Affairs of Chile hopes to promote the technological innovation of Chilean youth and international technological innovation, exchange and cooperation through Micromouse international competition, so as to promote the economic development of Chile. On December 3rd 2018, during the All Japan Micromouse International Competition, the Embassy of Chile in Japan hosted the "Chile International Micromouse Competition Seminar" and specially invited international experts (David Otten of the United States, Peter Harrison of the United Kingdom , Yukiko Nakagawa of Japan, Song Lihong of China, Benjamin of Chile,

etc.) jointly discussed the unified standards and specifications for the Micromouse maze international competition in Chile, as shown in Fig. A–9 and Fig. A–10.

Fig. A–9   Meeting of ministry of foreign affairs, Chile—discussion on the development of Micromouse

Fig. A–10   Chile Micromouse international competition seminar

6) Portuguese Micromouse International Competition

Portuguese Micromouse International Competition has helded in Gondomar on April 27, 2019, sponsored by University of Tras-os-Montes and Alto Douro, Portugal and the Technical Executive Committee.

Portuguese Micromouse International Competition, which started in 2011, aims to provide a complete technological learning environment through the cultivation of creativity and ability, and has been successfully held for 9 times.

Competition time: Every April/May.

Competition venue: Portugal.

The official website: https://www.micromouse.utad.pt/, as shown in Fig. A–11.

On April 27, 2019, at 18:00 local time, in Gondomar Coliseumm, Portual, teams from UK, China, Portugal, Spain, Brazil, Singapore and other countries, were competing a tense international Micromouse maze competition. With the steady search and fast sprint of Chinese Micromouse, applause and cheers

• Video

Portuguese
Micromouse
International
Competition

thundered in Gondomar Coliseumm...Micromouse of Qicheng achieved breakthrough results and won the second place in the world, as shown in Fig. A–12.

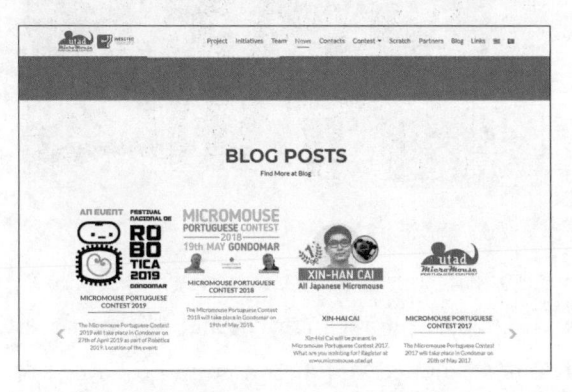

Fig. A–11   Screenshot of official website of Portuguese contest Micromouse International Competiton

Fig. A–12   Micromouse of Qicheng won the world second place in the Portugal competition

Antonio Valente, chairman of Micromouse Portuguese Contest Organizing Committee, said after the contest that in recent years, China's comprehensive national strength and technical strength have been increasing, especially in the field of education, more and more attention has been paid to technological innovation and engineering literacy. TQD-Micromouse participated in the Portugal international competition for the first time, and its excellent results are very gratifying, as shown in Fig. A–13.

Fig. A–13　On site technical exchange between Chinese and Portuguese Micromouse experts

7) India Interantional Micromouse Challenge

On January 4, 2020, the First International Micromouse Challenge of the 23rd Edition of Asia's Science and Technology Festival—Techfest 2020 was held in Mumbai, India. The delegations from India, China, Australia, Nepal, Sri Lanka, Bangladesh and other countries participated in the competition (see Fig. A–14). Micromouse team of Tianjin, China won all the medals of "gold, silver and copper" with absolute advantage, and successfully got the prize of 175,000 rupees.

Competition time: Every January.

Competition venue: Mumbai, India.

The official Website: http://techfest.org/competitions/Micromouse

● Video

India
Interantional
Micromouse
Challenge

Fig. A–14　Group photo of International Micromouse Challenge，India

It is particularly worth mentioning that the Luban Workshop team of Chennai Institute of technology in India(see Fig. A–15) that adopted the IEEE International standard equipment TQD-Micromouse-JD presented by China in 2017, won the champion of India domestic competition and the fourth place of

the world elite group and won the prize of 5,000 rupees, becoming the star team of International Micromouse Challenge, India. Kasik, teacher of Indian Luban workshop, said that such excellent results of the team of Luban Workshop in Chennai Institute of Technology is the result of the joint efforts of teachers and students of Luban Workshop and the support of enterprises, Tianjin Qicheng Science and Technology Co., Ltd. in the past three years.

Fig. A–15　Group photo of winners in the International Micromouse Challenge, India

8) Egypt International Micromouse Competition

Egypt IEEE Institute of Electrical and Electronics Engineers has developed into one of the most influential international academic and technical organizations now. For more than 30 years, the institute has been promoting and guiding the development and innovation of power electronics technology. This technology includes the effective use of electronic components, the application of circuit theory and design technology, and the development of analysis tools for effective conversion, control and power conditions. Our members include outstanding researchers, practitioners and outstanding prize-winners.

As shown in Fig.A–16, IEEE publicizes Micromouse competition on the home page of its official website. IEEE Conference on Power Electronics and Renewable Energy offers generous prizes for the winners of a high-profile international Micromouse competition. Grand Prize: An equivalent of $1,000, Outstanding Performance Award: An equivalent of $700, Best Innovative Design Award: an equivalent of $500. The teams are open to students from Egypt or international engineering or related majors in Egypt, as well as high school students. Within each group, a maximum of two students are allowed.

Fig. A–16　Egypt International Micromouse Competition

The official website: http://www.ieee-cpere.org/International_Competition.html.
Photos of Egypt International Micromouse Competition as shown in Fig. A–17.

Fig. A–17　Photos of Egypt International Micromouse Competition

Appendix B

Analysis of Improved-level Classic Competition Cases

IEEE International Invitational Competition is difficult, it is a challenging and interesting student competition that enjoys a certain reputation and influence at home and abroad. In terms of technology, the Micromouse competition covers Internet of things application technology, electronic information engineering technology, embedded technology, communication technology, software technology, computer network technology, information security technology, mobile communication technology, computer application technology, applied electronic technology, computer control technology, electromechanical integration technology, automation technology and other professional technologies. It involves skills and comprehensive professionalism in sensor detection, artificial intelligence, automatic control and electromechanical motion parts application and fully displaying the development of higher education and vocational education level and improves the training quality of high-quality and high skilled applied talents of electronic information.

The competition can promote the talent cultivation demand of electronic information industry, and adapt to the development of embedded technology, make the intelligent algorithm optimization and other advanced front-end technologies into the competition content in line with the development of science and technology, and guide electronic information majors to carry out single-chip microcomputer application, embedded technology application, Internet of things technology application, etc. The curriculum construction and teaching reform should promote the development of innovative personnel training mode, enhance the employment competitiveness of students majoring in electronic information, promote the education of innovation and entrepreneurship, strengthen the guidance and service of entrepreneurship, and improve the employment level.

With the development of technology, Micromouse conforms to the development of modern technology. After many years of transformation and

optimization, it has become an excellent practical education platform integrating new technologies such as artificial intelligence, embedded, and intelligent sensing.

For more than 40 years, IEEE has held an international Micromouse competition every year. Since its inception, students from all over the world have participated actively, especially college students in the United States and European countries. Therefore, some universities have also set up elective courses on "Micromouse Design Principles and Production Process", as shown in Fig B–1, Fig. B–2.

Fig. B–1   Professor David was in the competition

Fig. B–2   Masakazu Utsunomiya, Japan

Next, we will take typical maze maps of Micromouse competitions at home and abroad as examples to analysis.

1. Analysis of the 30th APEC International Micromouse Competition

On March 16, 2015, the 30th APEC international Micromouse competition was held successfully. Tianjin Qicheng Science and Technology Co.,Ltd. led a joint delegation of Tianjin university students to participate in the competition.

It is a historic turning point and the most significant milestone for Tianjin Micromouse technology to integrate with the world.

This competition is a very successful one. The sensor is transiting from digital type to analog type, and the motion structure is also developing from stepping motor to DC motor.

In this event, there is Micromouse integrating the ground suction fan.

The overall difficulty of the maze used in this competition is relatively balanced, and there are many paths to choose, including long straight path to show the high-speed movement performance, and moderate continuous turning to show the precise control, as shown in Fig. B–3.

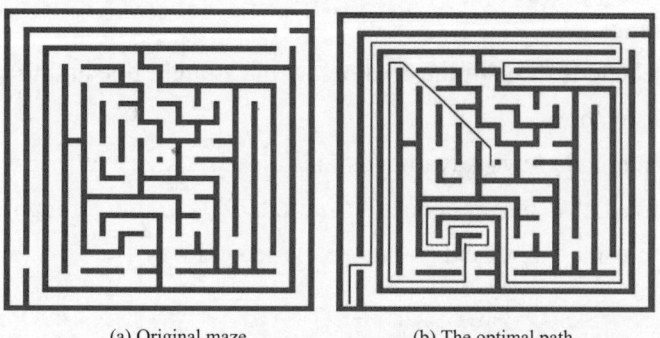

(a) Original maze   (b) The optimal path

(c) Crucial point A   (d) Crucial point B   (e) Crucial point C

Fig. B–3　Maze analysis

Crucial point A: It is consisting of three T-shaped structures. Combining different directions into more difficult graphics. Not only the accuracy of Micromouse sensor detection, but also the precise control of turning. Due to the large turning frequency and different turning directions, there is no time to correct the posture between turns; once Micromouse has a sensor detection error or a turning angle error, it will be difficult to pass successfully.

Crucial point B: After more than 40 years of development, Micromouse algorithms have been quite intelligent, they are all moving towards the end. Crucial point B is a falsely designed destination path. When Micromouse thinks this is the destination path, once it enters, it must pass through it smoothly. It is also difficult to combine multiple turns and concave paths. It is very important to test the accuracy of infrared detection and turning parameters of Micromouse.

Crucial point C: This is the only way to go to the destination. The difficulty lies in the continuous turning. There are two ways to choose in area C.

The first is to reach the end point through a diagonal line, it can choose to walk along the diagonal line at a 45° turn or turn continuously at 90° to finally reach the end point.

The second is through the area below the straight line and then 90° turning, the final destination.

Both methods have advantages and disadvantages. When walking at 45° can be realized, it is recommended to run the oblique line through the end point; otherwise, it is recommended to choose the second way.

2. Analysis of the "Qicheng Cup" Tianjin university student Micromouse competition

This event was hosted by Tianjin Municipal Education Commission, organized by Nankai University and Higher Education Committee of Tianjin Communication Society, and sponsored by Tianjin Qicheng Science and Technology Co.,Ltd. There are 150 teams from 20 colleges and universities participated in the competition, as shown in Fig. B-4.

Fig. B-4　Photo of the competition

Appendix | 101

The 6th "Qicheng Cup" College Students Micromouse Competition in Tianjin has been successfully held for seven times up to now. The Micromouse developed from a slow stepping motor and ultrasonic sensor to a coreless DC motor and a linear infrared sensor. The level of competitors is also increasing year by year. The 2017 "Qicheng Cup" Micromouse Competition features a classic maze, it is the most important characteristics of high difficulty and openness, as shown in Fig. B–5.

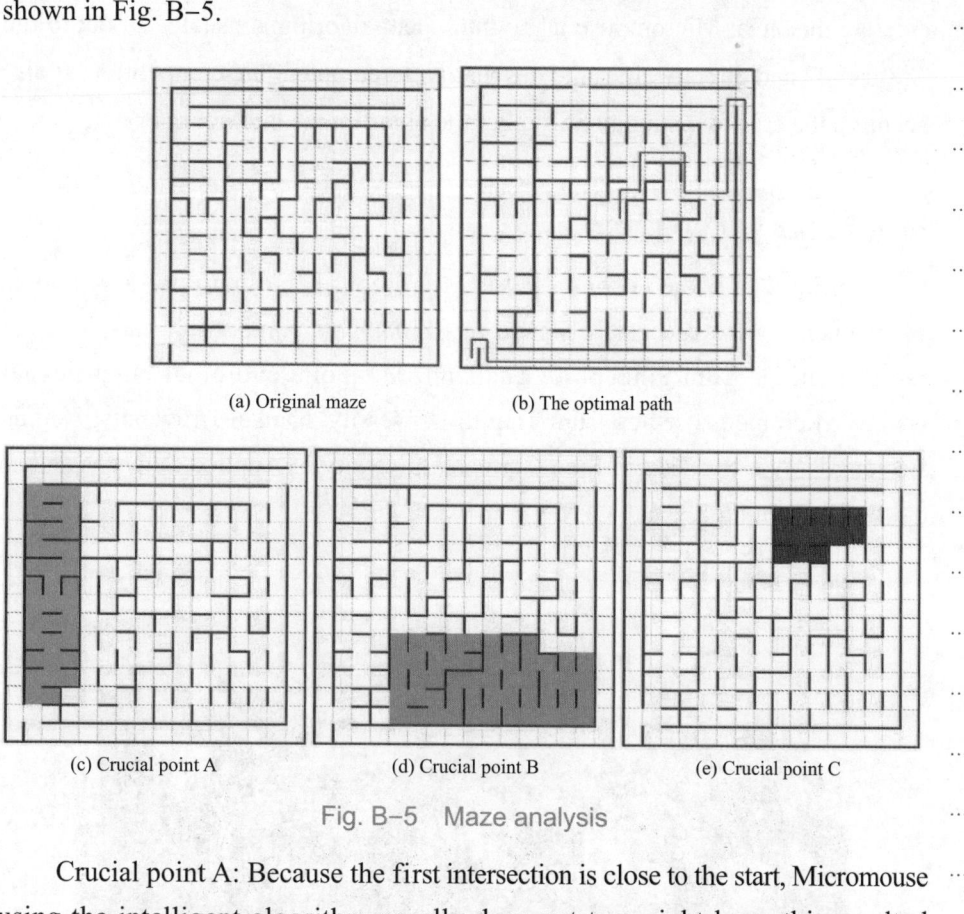

(a) Original maze   (b) The optimal path

(c) Crucial point A   (d) Crucial point B   (e) Crucial point C

Fig. B–5   Maze analysis

Crucial point A: Because the first intersection is close to the start, Micromouse using the intelligent algorithm usually does not turn right here, this resulted in the Micromouse walking all the way around the outermost wall and then coming back to the intersection and finally entering area A. Almost completely symmetrical patterns and numerous intersections make it almost impossible for Micromouse to correct their attitude. This is a great test for the performance of Micromouse.

Crucial point B: Compared with Crucial point A, area B is also highly open. There are so many intersections that Micromouse do not have enough time for attitude correction. At the same time, it gives high-performance Micromouse the opportunity to run 45° diagonals. It is easy to judge the level of competitors because of the performance gap of Micromouse.

Crucial point C: Area C is the only way to enter the destination. After entering this area, Micromouse using intelligent algorithms usually do not to run "dead end" and directly enter the destination to complete the competition. It also becomes the area to distinguish whether the algorithm is intelligent or not.

3. Analysis of the second IEEE Micromouse International Invitational Competition

In May 2017, the second "IEEE Micromouse International Invitational Tournament" were successfully held in China. This competition has attracted representatives from Singapore, Thailand, Mongolia and other countries, as well as Micromouse elites from Tianjin University, Nankai University, Beijing Jiaotong University, Tianjin Sino German University of Applied Sciences and so on, as shown in Fig. B-6.

Fig. B-6　Photo of the competition

This competition uses a more characteristic maze. With the popularity of intelligent algorithm, the maze selected for large-scale competition pays more and more attention to increase the difficulty coefficient for intelligent algorithm,

as shown in Fig. B–7.

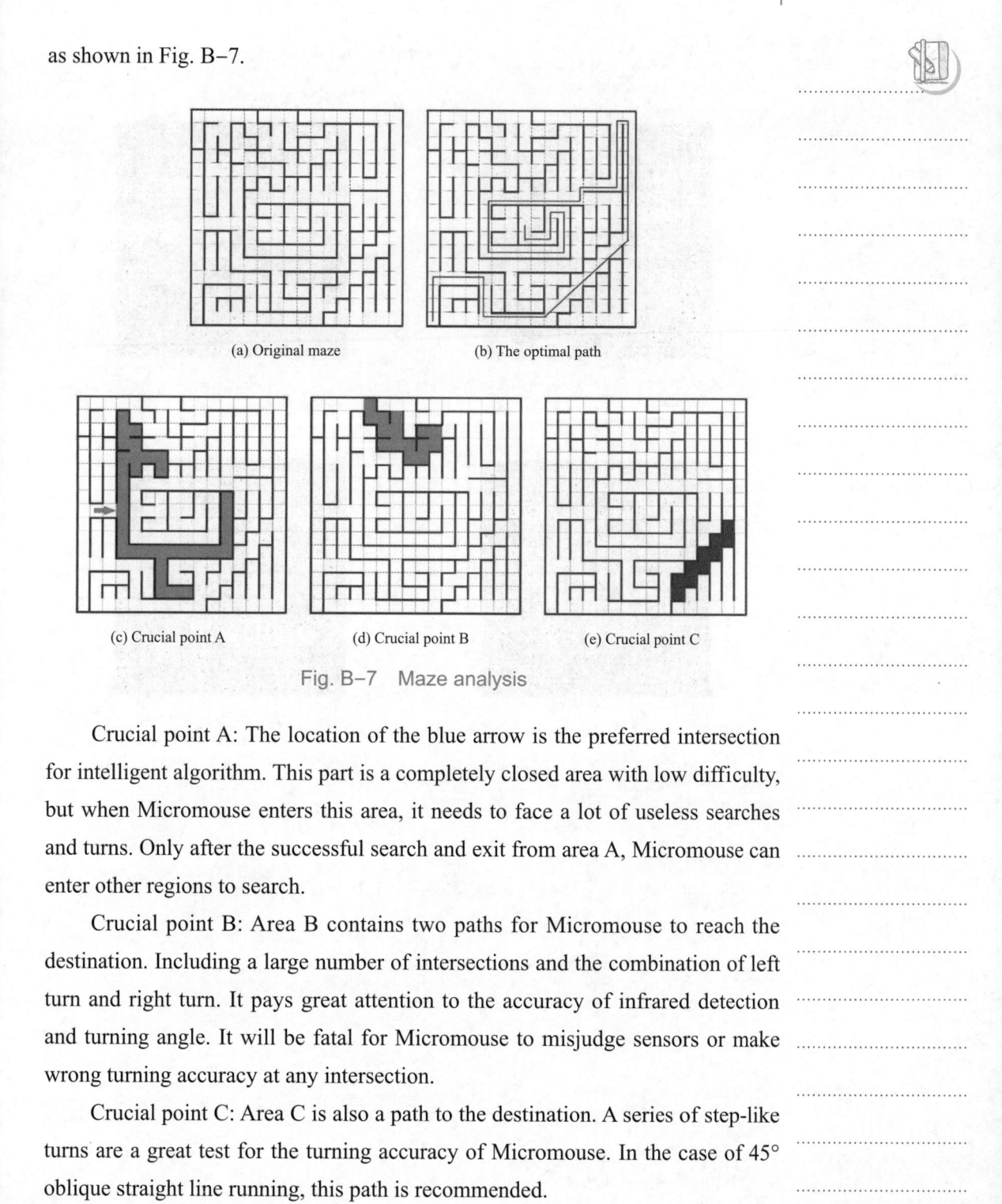

(a) Original maze   (b) The optimal path

(c) Crucial point A   (d) Crucial point B   (e) Crucial point C

Fig. B–7   Maze analysis

Crucial point A: The location of the blue arrow is the preferred intersection for intelligent algorithm. This part is a completely closed area with low difficulty, but when Micromouse enters this area, it needs to face a lot of useless searches and turns. Only after the successful search and exit from area A, Micromouse can enter other regions to search.

Crucial point B: Area B contains two paths for Micromouse to reach the destination. Including a large number of intersections and the combination of left turn and right turn. It pays great attention to the accuracy of infrared detection and turning angle. It will be fatal for Micromouse to misjudge sensors or make wrong turning accuracy at any intersection.

Crucial point C: Area C is also a path to the destination. A series of step-like turns are a great test for the turning accuracy of Micromouse. In the case of 45° oblique straight line running, this path is recommended.

4. Maze paradigms of International Micromouse Competition (see Fig. B-8-Fig. B-10)

Fig. B-8　All Japan Micromouse Contest,2012（expert competition and junior competition）

Fig. B-9　UK International Micromouse Invitational Competition, 2000

Fig. B-10　American APEC International Micromouse Competition, 2002

Appendix C

Device List of TQD-Micromouse-JD

Device list of TQD-Micromouse-JD is shown in Table C–1.

Table C–1　Device list of TQD-Micromouse-JD

No.	Name	Quantity	Remarks
1	TQD-Micromouse-JD	1	
2	Charger	1	
3	Downloader	1	
4	Connecting line	1	
5	USB line	1	
6	Battery	1	
7	Disk	1	

Appendix D

Teaching Content and Class Arrangement

The reference teaching hours are 48, and the allocation is shown in Table D–1.

Table D–1　Teaching content and class arrangement

No.	Teaching Contents		Class hours allocation
Chapter 1　Elementary Knowledge	Project 1	Evolution of Micromouse	20
	Project 2	Micromouse Hardware Structure	
	Project 3	Development Environment of Micromouse	
	Project 4	Basic Function Control of Micromouse	
Chapter 2　Comprehensive Practice	Project 1	Advanced Control Function	14
	Project 2	Actual Combat Tasks	
Chapter 3　Advanced Skills and Competitions	Project 1	Path Planning and Decision Algorithm	24
	Project 2	The Principle of Path Planning	
	Project 3	Micromouse Program Design	
Total			48

Appendix E

The Circuit Diagram Symbol Comparison Table

The circuit diagram symbol comparison table is shown in Table E–1.

Table E–1　The circuit diagram symbol comparison

No.	Name	Drawing methods under china national standard	Drawing methods in software
1	Light-emitting diode		
2	Diode		
3	Electrolytic capacitor		
4	Ground connection		
5	Button switch		
6	Adjustable resistor component		

Appendix 107

Appendix F

Bilingual Comparison Table of Glossary

Bilingual comparison table of glossary as shown in Table F–1-Table F–3.

Table F–1   Related to Micromouse

English	Chinese	English	Chinese
main control module	核心控制模块	the left, the front-left, the front, the front-right and the right	左方、左斜、前方、右斜、右方
main control chip	主控芯片	the g segment	g段
input module	输入模块	stepping motor	步进电动机
output module	输出模块	motor drive circuit	电动机驱动电路
main control circuit	核心板电路	truth table	真值表
power circuit	电源电路	H-bridge circuit	H桥电路
control circuit	控制电路	rotate	转动（步进电动机）
peripheral circuit	外围电路	electronic component	电子元器件
keyboard-display circuit	键盘显示电路	crystal oscillator	晶振
JTAG interface circuit	JTAG接口电路	capacitance	电容
key-pressingcircuit	按键电路	adjustable current-limiting resistance	限流可调电阻
data transmission	数据传输	digitron	数码管
human-computer interaction system	人机交互系统	peripheral device	外围器件
IR sensor	红外传感器	pulse oscillation circuit	脉冲振荡电路
infrared detection circuit	红外检测电路	pulse signal	脉冲信号
infrared light	红外线	square wave	方波
infrared calibration	红外校准	perceptual system	感知系统
infrared intensity	红外强度	carrier frequency	载波频率
infrared transmitter	红外发射头	schematic diagram	原理图
infrared receiver	红外接收头	software interface	软件界面
PWM signal generator driver module	PWM信号发生器模块	driver library	驱动库

Table F-2 Related to competition

English	Chinese	English	Chinese
cell	单元格	crossing	路口
wall	挡板	electronic automatic scoring system	电子自动计分系统
post	立柱	competitor	参赛队员
competition maze	竞赛场地	Micromouse competition	智能鼠竞速比赛
the start	起点	the optimal path	最优路径
the destination	目的地/终点	trajectory	轨迹
the coordinate in the maze	迷宫坐标	passage way	通道

Table F-3 Related to program

English	Chinese	English	Chinese
the bottom driver program	底层驱动	path planning and decision algorithm	路径规划和决策算法
the top algorithm program	顶层算法	struct	结构体
algorithm	算法	differential-speed control	差速控制
strategy	策略	straight movement	直线运动
rule	法则（左、右手法则）	turning	转弯
the right-hand rule，the left-hand rule，the central rule	右手、左手、中心法则	correct the attitude	校正车姿
90-degree turning/180-degree turning	90°、180°转弯	attitude correction	运行校正
programming and realizing	编程并实现	core function	核心函数
step map	等高图	time sequence status	（驱动步进电动机的）时序状态
cycle detection	循环检测	moving forward one cell	前进一格
movement control in picture-8-shaped path	"8字型"路径运行控制	waiting for button press	按键等待
obstacle avoidance	实现避障	determining the attitude	判断车姿
motion attitude control	运动姿态的控制	waiting one step	暂停一步
two-wheel difference speed	两轮差速	accurate turning control	精确转弯控制

Appendix | 109

Appendix G
The International Curriculum Standard for "Micromouse Design Principles and Production Process"

(Applicable to training courses in higher vocational colleges)

1. Applicable majors

Majors pertaining to electronic information, computer, communication and automation, etc.

Including: electronic information engineering technology (major code 610101), applied electronics technology (major code 610102), intelligent product development (major code 610104), intelligent terminal technology and application (major code 610105), electronic measurement technology and instrument (major code 610112), internet of things application technology (major code 610119), computer application technology (major code 610201), software technology (major code 610205), embedded technology and application (major code 610208), internet of things engineering technology (major code 610307), mechatronics technology (major code 560301), electrical automation technology (major code 560302), industrial process automation technology (major code 560303), intelligent control technology (major code 560304), industrial network technology (major code 560305), industrial automation instruments (major code 560306), industrial robot technology (major code 560309).

2. The orientation of the course

"Micromouse Design Principles and Production Process" training course is technically related to multiple technologies such as the Internet of things application, electronic information engineering, embedded technology, software technology, computer application technology, applied electronics technology, mechatronics technology, automation technology, intelligent control technology and so on, and covers skills and comprehensive professional capability such as sensor detection, artificial intelligence, automatic control and application of

electromechanical moving parts. In the training, we mainly study the hardware structure, the development environment, the infrared detection, the movement and attitude control, the path planning and behavior decision-making algorithm of Micromouse . It focuses on students' basic know-how in electrical and electronic technology, single-chip technology, motor control, electronic technology, embedded technology and programming, as well as the basic knowledge and capability related to engineering practice innovation.

The teaching content of this course adopts a project-based teaching model, and is arranged with increasing level of difficulty. In this way, a beginner is able to make progress from knowing a little about getting the hang of this course and finally, becoming a conversant learner. Learners can enrich practical engineering expertise and technical application experience. Their professional horizons are broadened, required professional qualities are cultivated and innovation abilities are improved in the process.

This course is offered in the first semester of the second grade. Its prerequisite courses are SCM Control Technology, C Language Programming, Electrical and Electronic Technology and Sensors and Detection Technology and its subsequent courses are Automatic Control Technology and Object-Oriented Programming.

3. Course objectives

This course regards the following aspects as the core of students' professional competencies: the hardware structure, the development environment, the infrared detection, the movement and attitude control, the path planning and behavior decision-making algorithm of Micromouse . The method combining "teaching, learning and experimenting" is adopted in order to build students' professional competences, social ability and methodological ability.

1) Professional competences

(1) Master the use of common instruments.

(2) Grasp the basic principles and methods of Micromouse assembly and debugging.

(3) Grasp the basic usage of embedded development environment.

(4) Master the method of programming the overall project.

(5) Get the hang of analyzing and processing sensor signals.

(6) Master the method of stepper motor control.

(7) Master the relevant knowledge of intelligent search and path planning in a maze.

(8) Able to analyze common breakdowns and provide troubleshooting methods during the Micromouse device debugging.

2) Methodological ability

(1) Able to independently collect and sorting related data and information.

(2) Able to independently formulate and implement work plans.

(3) Have the ability to explore and use theoretical knowledge to solve real-life problems.

3) Social ability

(1) Ability of communition and teamwork

(2) Be innovative, enthusiastic and devoted in work.

(3) Have an awareness of safety, quality and responsibility.

4. Design framework

A total of 5 projects are designed. In the project, the task objectives, task content, knowledge and ability requirements for teachers, knowledge and ability preparation requirements for students, teaching materials, implementation steps, and the number of class hours required to complete the project are made clear. A six-step method following an order of introduction→ collection → target setting → implementation → experiment → assessment is adopted in teaching organizing.

Class hours for reference: 48 hours. Total credits: 3.

5. Content outline

(1) Targeted group: students of three-year tertiary vocational programs.

(2) Class hours for reference: 48 hours.

(3) Learning objectives: Learners will learn the analysis and assembly of the hardware system structure of Micromouse , the building of the developing environment, sensor and detection signals commissioning and be able to realize the forward-moving and turning control and pose correction, intelligent search and path planning.

(4) Work projects:

Project 1   The hardware structure of Micromouse (6 class hours)

Project description: Preliminary understanding of the structure of the TQD-Micromouse-JD. Analyzing the mechanical structure design principle, understanding the electronic components of the TQD-Micromouse-JD, their working principle, features and installation methods, and completing the Micromouse hardware assembly.

Project objectives: (For learners) to master the functions and features of the core controller. To familiarize with the basic common instrument testing methods, and to try to measure related circuit modules (circuit on / off, VCC, GND, etc.). To learn the assembly of the TQD-Micromouse-JD under the guidance of manuals or related drawings.

Project 2   The development environment of Micromouse (4 class hours)

Project description: Familiarize learners with the installation and use of IAR, an international open source software for Mircomouse's development environment. Enable learners to debug the program via IAR, and download the program for Micromouse .

Project objectives: Familiarize learners with the methods of software installation and operation, and installation of related drivers. Enable students to properly connect the TQD-Micromouse-JD and downloader with a computer and download the program.

Project 3   The infrared detection of Micromouse (10 class hours)

Project description: Learn about the working principle and use methods of the sensors on the TQD-Micromouse-JD device. Understand the role of the digitron of the expanded module and buttons.

Project objectives: Learn the usage of keys and digitrons. Master the distribution principle of Micromouse device sensors on circuit boards. Through the commissioning process, learn to use sensors to measure distances and display through digitrons to achieve multi-sensor collaboration.

Project 4   The movement and attitude control of Micromouse (14 class hours)

Project description: Understanding the motion structure of TQD-Micromouse-JD—the control principle of stepper motor, and being able to control the forward-moving and turning of Micromouse .

Project objectives: Master the driving method of the Micromouse stepping motor, and learn how to adjust speed and turning angle. Learn how to correct Micromouse movements through sensors.

Project 5   The path planning and behavior decision-making algorithm of Micromouse (14 class hours)

Project description: Learn about the algorithms relating to intelligent search and path planning of Micromouse to realize path planning in mazes.

Project objectives: Learn the definition of the maze coordinates and the storage method of the wall information, learn the path planning method for Micromouse in a maze—map step edit, have a deft use of the left-hand rule, right-hand rule, and the center rule to search and memorize the maze, learn the program structure of Micromouse competition, and work out the optimal path.

6. Skill assessment requirements

Assessment will be conducted in both theory and practice.

The overall assessment results are calculated on a 100-point scale and include two parts, the usual performance assessment plus the final comprehensive assessment. The usual performance assessment accounts for 30% of the total, including attendance, project works, and tests; the final comprehensive assessment accounts for 70% of the total, including 30% of theoretical exams and 40% of practical operation.

7. Implementation suggestions

(1) Teachers should arrange and organize teaching activities based on typical products in the work tasks.

(2) Teachers should compile project assignments in accordance with the project's learning objectives. Each assignment should specify the content of the teacher's teaching (or demonstration); clarify the requirements of the learner's preview; propose the overall arrangement of the project and the time and content of training for each module. If students are asked to study in groups, then requirements should be specified as to the division of groups and the requirements for group discussion (or operation).

(3) Teachers should follow a learner-centered approach in designing teaching syllabus so as to create a democratic and harmonious teaching ambiance,

stimulate learners to participate in teaching activities, increase their enthusiasm for learning, and enhance their confidence and sense of achievement.

(4) Teachers should guide learners to finish the project completely and integrate expertise, skills, professional ethics and emotional attitudes in the process.

8. Teaching conditions

Micromouse innovation maker training rooms equipped with professional textbooks that can guide teaching.

9. Learning evaluation

Teacher evaluation: to evaluate and give feedback to students in every step of practice.

Student evaluation: after the students finish the tasks in groups, two class hours will be spent in learning from each other and giving mutual evaluation so as to improve their performance next time.

10. Teaching materials and reference materials

1) Recommended teaching material

王超 高艺 宋立红. 智能鼠原理与制作：进阶篇[M]. 北京：中国铁道出版社有限公司，2019.

2) Reference materials

［1］黄智伟. 32位ARM微控制器系统设计与实践[M]. 北京：北京航空航天大学出版社, 2010.

［2］来清民，来俊鹏. ARM Cortex-M3嵌入式系统设计和典型实例[M]. 北京：北京航空航天大学出版社, 2013.

11. Remarks

Editor: Yao Song, Liu Baosheng, Song Lihong, Fan Pingping, Li Meng, Chen Likao, Qiu Jianguo

Reviewer: Gong wei, Gao yi, Kang xiaoming

March 24, 2020